Biomedical Image Analysis: Segmentation

Biomedical Image Analysis: Segmentation
Scott T. Acton and Nilanjan Ray

ISBN: 978-3-031-01117-7 paperback

ISBN: 978-3-031-02245-6 ebook

DOI: 10.1007/978-3-031-02245-6

A Publication in the Springer series

SYNTHESIS LECTURES ON IMAGE, VIDEO AND MULTIMEDIA PROCESSING #9

Lecture #9

Series Editor: Alan C. Bovik, University of Texas, Austin

Series ISSN
ISSN: 1559-8136   print
ISSN: 1559-8144   electronic

Biomedical Image Analysis: Segmentation

Scott T. Acton
University of Virginia

Nilanjan Ray
University of Alberta

SYNTHESIS LECTURES ON IMAGE, VIDEO AND MULTIMEDIA PROCESSING #9

ABSTRACT

The sequel to the popular lecture book entitled *Biomedical Image Analysis: Tracking*, this book on *Biomedical Image Analysis: Segmentation* tackles the challenging task of segmenting biological and medical images. The problem of partitioning multidimensional biomedical data into meaningful regions is perhaps the main roadblock in the automation of biomedical image analysis. Whether the modality of choice is MRI, PET, ultrasound, SPECT, CT, or one of a myriad of microscopy platforms, image segmentation is a vital step in analyzing the constituent biological or medical targets. This book provides a state-of-the-art, comprehensive look at biomedical image segmentation that is accessible to well-equipped undergraduates, graduate students, and research professionals in the biology, biomedical, medical, and engineering fields. Active model methods that have emerged in the last few years are a focus of the book, including parametric active contour and active surface models, active shape models, and geometric active contours that adapt to the image topology. Additionally, *Biomedical Image Analysis: Segmentation* details attractive new methods that use graph theory in segmentation of biomedical imagery. Finally, the use of exciting new scale space tools in biomedical image analysis is reported.

KEYWORDS

image segmentation, computer vision, image processing, medical imaging, biological imaging, active contours, graph theory, scale space, level set analysis

Dedication

To Henry Harper Acton
To Nilarnab Ray

Contents

CHAPTER 1

Introduction

Isn't it strange that I who have written only unpopular books should be such a popular fellow?

—Albert Einstein

We cannot presume to write a book that addresses all of biomedical image segmentation. What we have done is to write a book that covers our recent developments in the area and those developments to which we have been exposed through practical experience.

What is segmentation? According to Euclid, the whole is defined by the sum of the parts. In image segmentation, we attempt to find the parts. In biomedical image segmentation, these parts may represent cells or molecules, organs or tissues, to name a few. The problem of biomedical image segmentation is daunting.

Coastlines, doorways, integrated circuits, printed text, tanks. They share in common rigid boundaries.

Cells morph, split, roll, tether, spiculate, merge, engulf, ingest, transmigrate, tear, transport, differentiate, replicate, detect, adhere, protect, grow. Organs beat, pump, break, remodel, fail, respire, transmit, react, perceive, clean, regulate, filter, metabolize, breath, think.

Such is the contrast of biomedical image analysis with the other prime targets of image analysis and computer vision over the last 35 years. Biomedical targets are typically targets in action, deformable, difficult.

It could be argued that of all vital biomedical imaging problems, we have worked the most and achieved the least in segmentation. Fundamental image reconstruction algorithms from magnetic resonance to computer-aided tomography are well established. These algorithms are efficient and implemented in common off-the-shelf hardware and software. Visualization and computer graphics for medical imaging have followed computer gaming into commodity products. Progress is still being made in multimodal and longitudinal registration, but several established techniques exist and are utilized by the major medical imaging companies. Enhancement and restoration algorithms have all but stabilized in the industry, although specialized algorithms will continue to be developed.

Commodity, *off-the-shelf*, and *accepted solution* are not words that come to mind when discussing segmentation. Largely, although tools exist in many forms of biomedical imaging products, image segmentation remains in the hands of the research community. It is neither standardized nor, in general, efficient.

It remains to be seen if segmentation software will become a commodity product. We would contend that successful and efficient algorithms for image segmentation are critical to the future of biomedical image analysis. How can the richness of cellular images from high-content screening be harvested without effective, fast segmentation? How can the four-dimensional information from new, powerful magnetic resonance platforms be used without segmentation? How can computer-assisted surgery become a ubiquitous tool without segmentation?

Our view is that the realization of image segmentation as a widespread commercial tool will be maturing in the next decades. The algorithms and techniques presented here do not yet represent maturity. In fact, the algorithms currently represent a roadblock—a bottleneck—in the image analysis pipeline. We are excited to be part of the early era in biomedical image segmentation. As the use of open-source tools, commercial tools, and shared research tools in segmentation becomes more commonplace, we anticipate significant advances in areas such as systems biology and in clinical practice itself.

The remainder of the book consists of five additional chapters. Chapter 2 covers parametric active contours and their application in biomedicine. Special emphasis is given to the evolution procedure, the external force models, and the parameterization. In Chapter 3 on Bayesian active contour methods, we shift our focus to active shape models and other deformable models that take a training set or other prior information into account. Geometric active contours for biomedical segmentation are covered in Chapter 4. These special contours can split and merge, accommodating a wide range of topologies. In Chapter 5, we explore new graph theoretic methods of segmentation that have emerged recently in the biomedical community. The final chapter, Chapter 6, treats methods to segment and prefilter images using scale space.

Scott Acton, Santa Fe, New Mexico USA
Nilanjan Ray, Edmonton, Alberta, Canada

CHAPTER 2

Parametric Active Contours

Do you want to know who you are? Don't ask. Act! Action will delineate and define you.

—Thomas Jefferson

2.1 OVERVIEW

One of the revolutionary computational models in image segmentation was introduced in the late eighties. It is known as snakes or active contours. According to Google Scholar, the "snake" paper published by M. Kass, A. Witkin, and D. Terzopoulos in 1988 in the *International Journal of Computer Vision* has been cited more than 5,000 times to date! In biomedical image analysis, snake methods have created a tremendous sensation. There are quite a few reasons behind this huge success story of snakes. During the course of this chapter, we unravel the snake evolution methods, examine the factors behind the popularity, and illustrate different biomedical applications for snakes.

2.2 WHAT IS A PARAMETRIC ACTIVE CONTOUR?

During the time in which the Human Genome Project received wide publicity, Kass et al. [1] had quietly let the snake loose. Ever since, the snake (aka active contour) has become one of the most popular object delineation methods in image analysis, particularly in biomedical and medical image analysis. An active contour is a thin elastic band that is placed on an image in anticipation that it will delineate the desired object.

To study the basic snake model introduced by Kass et al., we need to understand the parametric curve. A continuous curve on a two-dimensional plane can be defined by its x and y coordinates, where these coordinate values are continuous functions of a scalar parameter, say s. Thus, we can say that $(X(s), Y(s))$ represent a two-dimensional continuous curve where $X(s)$ and $Y(s)$ are two continuous functions of s representing, respectively, the x and the y coordinate values. The scalar parameter s is often allowed to take values between 0 and 1, i.e., $s \in [0, 1]$. $(X(0), Y(0))$ represents one end of the curve and $(X(1), Y(1))$ the other end. So, how is a closed curve

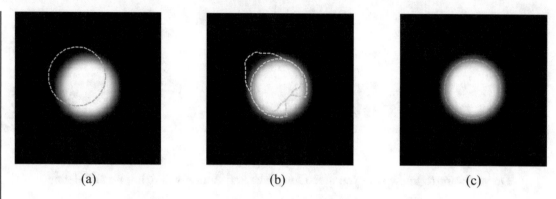

     (a)   (b)   (c)

FIGURE 2.1: (a) An image with a circular object. The initial snake is overlaid here. (b) Two intermediate snake locations corresponding to the end of two different iterations during snake evolution are shown. (c) Final snake position delineating the circular object.

represented? Simply define $(X(0), Y(0))$ and $(X(1), Y(1))$ as the same point, i.e., $X(0) = X(1)$ and $Y(0) = Y(1)$. As for example, $X(s) = \cos(2\pi s)$ and $Y(s) = \sin(2\pi s)$, $s \in [0, 1]$ defines a closed curve: a circle with unit radius and its center at the origin.

    The goal of the basic snake model is to delineate the boundary of an object of our interest. To achieve this segmentation, the snake is first placed near the object boundary. This step is sometimes referred to as *snake initialization*. See Figure 2.1a showing an object (a circle in this case) and an initial snake. Next, a computational process called *snake evolution* is performed. Snake evolution is an iterative computation where the snake is designed to slither across an image. Figure 2.1b illustrates a snake evolution. A good snake evolution will push the initial snake toward the object boundary and once the snake reaches the object boundary, any further snake evolution makes no changes. We say that the snake evolution has converged when the evolution ceases. Figure 2.1c illustrates the final snake after evolution has stopped/converged.

    Various questions at this point must arise within curious minds, such as how is the snake initialization performed—automatically or with the help of the user? How close should the snake be placed to the object boundary in order to capture the object? How is the snake evolution computation performed? And so on. The next sections treat these issues.

2.3   ACTIVE CONTOUR EVOLUTION

Active contour evolution can be explained very conveniently once we pay some attention to the gradient magnitude of an image. Figure 2.2a shows the image $I(x,y)$ of Figure 2.1a as a surface plot. Figure 2.2b shows negative of $f(x,y)$, where $f(x,y) = |\nabla I|^2 = (\partial I/\partial x)^2 + (\partial I/\partial y)^2$, i.e., the gradient magnitude squared of the surface/image $I(x,y)$. We notice that the snake essentially crawls down

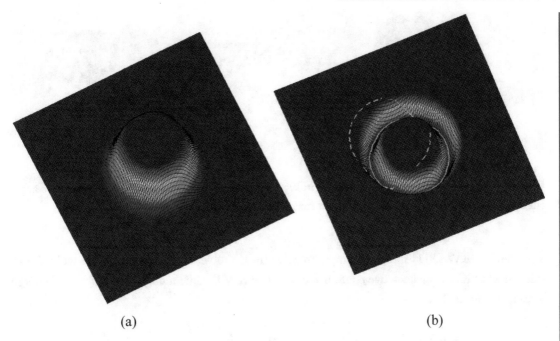

(a)   (b)

FIGURE 2.2: (a) Surface plot for the image in Figure 2.1a. (b) Negative squared gradient magnitude of the surface shown in (a). The initial and the final snake locations are also overlaid.

the *negative of squared gradient surface* (-*f*) and settles at its valley. Mathematically, we say that snake evolution is attempting to minimize the following energy functional:

$$E_{\text{ext}}(X, Y) = - \int_0^1 f(X(s), Y(s)) \, ds. \qquad (2.1)$$

Notice the use of the word *functional* here. Because E_{ext} is a function of X and Y that are themselves functions of s. The E term with the subscript "ext" stands for *external energy*, which is a function of the entire image I.

Deviations from smoothness in the gradient magnitude surface cause problems for the snake evolution process. Figure 2.3a depicts the same circular object of Figure 2.1a with zero mean Gaussian noise added to it. Note the small undulations present in the surface. These are the effects of Gaussian noise. Gaussian noise is a good approximation to the noise introduced by a majority of imaging sensors into the real-life images. Figure 2.3b shows the associated negative gradient surface. The active contour cannot reach the valley after its evolution converges (see Figure 2.4a). As a consequence, the segmentation result is also flawed. Moreover, the final active contour is also not very smooth. Is there any way to resolve these issues? The basic snake model of Kass, Witkin, and

(a)　　　　　　　　　　　　　　　　(b)

FIGURE 2.3: (a) Surface plot for the image in Figure 2.1a after adding uncorrelated zero mean Gaussian noise at it. (b) Negative squared gradient magnitude of the surface shown in (a).

Terzopoulos (KWT) [1] indeed attempts to solve these problems (from now on, we will refer to this model as KWT snake/active contour model). The KWT snake model adds an internal energy term to Equation 2.1:

$$E_{\text{int}}(X, Y) = \frac{1}{2} \int_0^1 \alpha \left[\left| \frac{dX}{ds} \right|^2 + \left| \frac{dY}{ds} \right|^2 \right] + \beta \left[\left| \frac{d^2X}{ds^2} \right|^2 + \left| \frac{d^2Y}{ds^2} \right|^2 \right] ds, \qquad (2.2)$$

where α and β are some tunable parameters. Sometimes, these tunable parameter values can be computed automatically using minimax principle (e.g., see Reference [2]). As already mentioned, this internal energy should make the snake smooth, taut, but not jagged. Why? In reality, the object

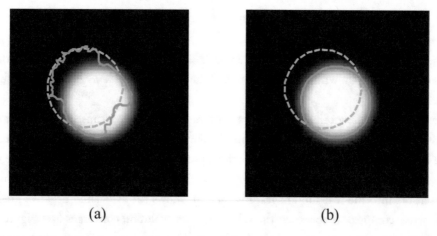

(a)　　　　　　　　　　　　　　　　(b)

FIGURE 2.4: (a) Due to noise, the snake evolution fails to delineate the circle. (b) After adding regulation terms, the snake is able to delineate the circle.

boundaries are often smooth, or at least piecewise smooth, i.e., typically there are only a handful of sharp corners. Let us see now why the internal energy term encourages a smooth active contour. Note that a contour with lower internal energy should have average lower magnitudes, the first and the second derivative of X and Y with respect to the parameter s. So what do the quantities $(dX/ds, dY/ds)$ and $(d^2 X/ds^2, d^2Y/ds^2)$ mean? $(dX/ds, dY/ds)$ roughly indicates the change we get in the coordinate locations of the curve as a result of unit change in the parameter s. Obviously, if a curve is rough, this change would be large. In fact, if we think of s denoting time, then these two quantities represent the velocity and the acceleration with which a particle is traveling along the contour. So if we want a smooth curve, or in other words, a gentle motion of the particle, we would need to limit the magnitude of the velocity and the acceleration. With this rationale, the basic snake model combines the internal and the external energy into:

$$E_{total}(X,Y) = E_{int}(X,Y) + E_{ext}(X,Y) \qquad (2.3)$$

and minimizes it with respect to $X(s)$ and $Y(s)$. Does this kind of imposed smoothness avoid the effect of noise on an image? The answer is "yes," to a reasonable extent. Figure 2.4b illustrates that, after addition of the internal energy component, the active contour is able to overcome the impediments or bumps created by Gaussian noise.

Time has come to reveal the oracle of snake evolution. In order for evolution computation, one needs to minimize the energy (Equation 2.3). However, note that Equation 2.3 is a functional, i.e., a function of a function. For these cases, we need an advanced optimization tool from functional analysis, known as calculus of variations or variational calculus [3]. This tool tells us how to take a derivative with respect to a function. After taking the functional derivative, one would follow essentially the high school calculus procedure—equate the derivative to zero and solve the equation for finding the minimum. The functional derivatives of the energy functional (Equation 2.3) equated to zeros are provided here:

$$\frac{\delta E}{\delta X} = -\alpha \frac{d^2 X}{ds^2} + \beta \frac{d^4 X}{ds^2} - \frac{\partial f}{\partial x} = 0, \qquad (2.4)$$

and

$$\frac{\delta E}{\delta Y} = -\alpha \frac{d^2 Y}{ds^2} + \beta \frac{d^4 Y}{ds^2} - \frac{\partial f}{\partial y} = 0. \qquad (2.5)$$

These two partial differential equations are known as *Euler equations*. Closed form solutions for X and Y from Equations 2.4 and 2.5 cannot be obtained in general. We resort to numerical techniques. One well-known method for solving Equations 2.4 and 2.5 is called the gradient descent method.

We start with a simple example to elucidate the gradient descent procedure. Figure 2.5 shows a function: $h(x) = x^4 + x^2 + 4x + 1$. We want to find the value of x at which $h(x)$ is minimum. Let us

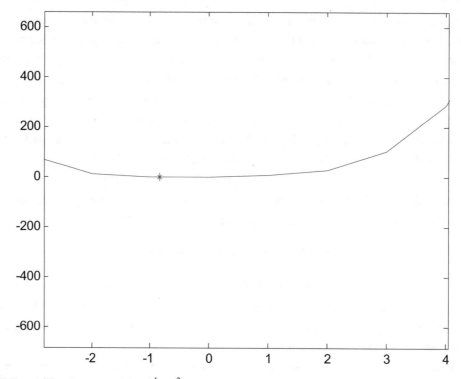

FIGURE 2.5: The function $h(x) = x^4 + x^2 + 4x + 1$ plotted within an x range of -3 to $+4$. The minimum point is at -0.8351 shown by an asterisk.

call this value of x as x^*. From high school calculus lessons, we know that x^* is a solution of the equation: $dh(x)/dx = 4x^3 + 2x + 4 = 0$. Finding x^* may not be an easy task algebraically. One way might be to use the principle of gradient descent here. With the gradient descent principle, we start with an initial guess solution value, say $x = 0$. Then, we change the value of the variable x iteratively (i.e., repeatedly), so that it approaches x^* gradually.

So, now the question is how to change the value of x. The gradient descent rule says that the change in x should be proportional to the negative of the derivative of $h(x)$. Intuitively, this makes sense, once we look at Figure 2.5. For example, at $x = 0$, the value of the derivative $dh(x = 0)/dx$ is positive, so we should decrease the current value of x that is 0 in this case. In order to reach minimum point x^*, we apply this gradient descent rule repeatedly. Take another example, say, if we are at $x = -1$, $dh(x = -1)/dx$ is negative. Thus, we should increase the current value. Mathematically, this gradient descent rule can be said as $\dfrac{dx}{d\tau} = -\dfrac{dh(x)}{dx}$, where τ denotes a pseudo-time variable. In other words, the instantaneous change in the variable x is equal to the negative of the derivative of the function h. This is the golden rule of gradient descent. Interestingly, note that when the minimum value is reached, the derivative value is zero, and hence, there is no need to make any changes to the x value! For the sake of illustration, we said here that to solve the equation $dh(x)/dx = 0$, one may apply gradient descent. In practice, however, one would want to apply a more efficient technique know as Newton's method here. With Newton, we find that the solution is $x^* = -0.8351$. It turns out that for snake evolution, Newton's method is computationally much more expensive than gradient descent.

From our discussions about the gradient descent method, it may be apparent that to solve the Euler Equations 2.4 and 2.5, we apply the following gradient descent rules:

$$\frac{\partial X}{\partial \tau} = \alpha \frac{\partial^2 X}{\partial s^2} - \beta \frac{\partial^4 X}{\partial s^2} + \frac{\partial f}{\partial x}, \tag{2.6}$$

and

$$\frac{\partial Y}{\partial \tau} = \alpha \frac{\partial^2 Y}{\partial s^2} - \beta \frac{\partial^4 Y}{\partial s^2} + \frac{\partial f}{\partial y}. \tag{2.7}$$

Equations 2.6 and 2.7 can be interpreted as force balance equations that drive the snake. In these equations, the internal force (the stretching and the bending terms) and the external force (terms involving image gradients) are trying to balance each other. When the net result force is zero, the snake stops evolving, and we obtain a local minimum of Equation 2.3.

The snake evolution is performed as follows. We start with initial snake configurations, i.e., with an initial $X(s)$ and $Y(s)$. However, to represent the snake, we need a polygonal representation of the continuous contour. This is performed by representing the continuous parameter $s \in [0, 1]$ with indices $i \in \{0, 1,..., n - 1\}$, with n being the total number of snaxels. Thus, the discrete counterpart of $(X(s), Y(s))$ is (X_i, Y_i).

The discrete forms for the Equations 2.6 and 2.7 are as follows:

$$\frac{X_i^{\tau+1} - X_i^\tau}{\zeta} = \alpha(X_{i+1}^\tau - 2X_i^\tau + X_{i-1}^\tau)$$
$$- \beta(X_{i+2}^\tau - 4X_{i+1}^\tau + 6X_i^\tau - 4X_{i-1}^\tau + X_{i-2}^\tau) + f_x(X_i^\tau, Y_i^\tau), \tag{2.8}$$

and

$$\frac{Y_i^{\tau+1} - Y_i^\tau}{\zeta} = \alpha(Y_{i+1}^\tau - 2Y_i^\tau + Y_{i-1}^\tau)$$
$$- \beta(Y_{i+2}^\tau - 4Y_{i+1}^\tau + 6Y_i^\tau - 4Y_{i-1}^\tau + Y_{i-2}^\tau) + f_y(X_i^\tau, Y_i^\tau), \tag{2.9}$$

where τ and $\tau + 1$ represent two successive time instants with step length ζ. The $+$ and $-$ operations involving the subscripts of X and Y in Equations 2.8 and 2.9 are modulo n operations. For example, the addition $(n - 1) + 2$ yields 1, as opposed to $n + 1$. These modulo operations take into account the wraparound for a closed contour. In Equations 2.8 and 2.9, we have used second- and fourth-order approximations, respectively, for the first and second derivatives. For better accuracy, one can use a fourth-order approximation for the first derivative, such as, $-(1/12)X_{i-2} + (4/3)X_{i-1} - (5/2)X_i + (4/3)X_{i+1} - (1/12)X_{i+2}$.

The edge-based force f_x and f_y in Equations 2.8 and 2.9 are as follows:

$$f_x(x,y) = \frac{\partial}{\partial x} f(x,y) \tag{2.10}$$

and

$$f_y(x,y) = \frac{\partial}{\partial y} f(x,y). \tag{2.11}$$

Sometimes, the functions $(f_x(x,y), f_y(x,y))$, or its shorthand notation (f_x, f_y), are referred to as a vector field that acts as *external forces* for the snake. It is convenient to express Equations 2.10–2.11 in matrix–vector notations. In order to do that, we employ the following notation:

$$\mathbf{x}^{\tau} \equiv [X_0^{\tau},\ldots, X_{n-1}^{\tau}]^{\mathrm{T}}, \tag{2.12}$$

$$\mathbf{y}^{\tau} \equiv [Y_0^{\tau},\ldots, Y_{n-1}^{\tau}]^{\mathrm{T}}, \tag{2.13}$$

$$\mathbf{f}_x^{\tau} \equiv [f_x(X_0^{\tau},Y_0^{\tau}),\ldots, f_x(X_{n-1}^{\tau},Y_{n-1}^{\tau})]^{\mathrm{T}}, \tag{2.14}$$

and

$$\mathbf{f}_y^{\tau} \equiv [f_y(X_0^{\tau},Y_0^{\tau}),\ldots, f_y(X_{n-1}^{\tau},Y_{n-1}^{\tau})]^{\mathrm{T}}. \tag{2.15}$$

Using this notation, Equation 2.8 can be written as:

$$\frac{\mathbf{x}^{\tau+1} - \mathbf{x}^{\tau}}{\zeta} = -A\mathbf{x}^{\tau} + \mathbf{f}_x^{\tau}, \tag{2.16}$$

and similarly, Equation 2.9 can also be written as:

$$\frac{\mathbf{y}^{\tau+1} - \mathbf{y}^{\tau}}{\zeta} = -A\mathbf{y}^{\tau} + \mathbf{f}_y^{\tau}. \tag{2.17}$$

The matrix A in Equations 2.16 and 2.17 is of size n-by-n with an almost pentadiagonal structure:

$$A = \begin{bmatrix} c & b & a & & & & a & b \\ b & c & b & a & & & & a \\ a & b & c & b & a & & & \\ & \ddots & \ddots & \ddots & \ddots & \ddots & & \\ & & a & b & c & b & a \\ a & & & & a & b & c & b \\ b & a & & & & a & b & c \end{bmatrix}, \tag{2.18}$$

where a, b, and c are as follows:

$$a = \beta, \quad b = -(4\beta + \alpha), \quad c = 6\beta + 2\alpha. \tag{2.19}$$

To iteratively solve for the snake location (the set of snaxel positions), we rewrite Equations 2.16 and 2.17, respectively, as:

$$\mathbf{x}^{\tau+1} = \mathbf{x}^{\tau} - (\zeta)A\mathbf{x}^{\tau} + (\zeta)\mathbf{f}_x^{\tau}, \tag{2.20}$$

and

$$\mathbf{y}^{\tau+1} = \mathbf{y}^{\tau} - (\zeta)A\mathbf{y}^{\tau} + (\zeta)\mathbf{f}_y^{\tau}. \tag{2.21}$$

Equations 2.20 and 2.21 are known as *explicit* solution techniques [4]. The numerical stability of Equations 2.20 and 2.21 depends on the time step ζ. For practical time step values, Equations 2.20–2.21 may be unstable. Instead, a *semi-implicit* procedure [4] is followed by rewriting them as:

$$\frac{\mathbf{x}^{\tau+1} - \mathbf{x}^{\tau}}{\delta\tau} = -A\mathbf{x}^{\tau+1} + \mathbf{f}_x^{\tau}, \tag{2.22}$$

and

$$\frac{\mathbf{y}^{\tau+1} - \mathbf{y}^{\tau}}{\delta\tau} = -A\mathbf{y}^{\tau+1} + \mathbf{f}_y^{\tau}. \tag{2.23}$$

Note that in the semi-implicit method, a backward time difference is made for the left-hand side of the equations. However, the entire right-hand side could not be written at the same time point because data term is only available at the current time point. From Equations 2.22–2.23, we immediately obtain the following iterative forms:

$$\mathbf{x}^{\tau+1} = (I_n + (\zeta)A)^{-1}(\mathbf{x}^{\tau} + (\zeta)\mathbf{f}_x^{\tau}), \tag{2.24}$$

and

$$\mathbf{y}^{\tau+1} = (I_n + (\zeta)A)^{-1}(\mathbf{y}^{\tau} + (\zeta)\mathbf{f}_y^{\tau}). \tag{2.25}$$

where I_n is the n-by-n identity matrix. From elementary linear algebra, it can be shown that the matrix $(I_n + (\zeta)A)$ is invertible. Equations 2.24 and 2.25 are sometimes referred to as the snake evolution equations. Algorithm KWT illustrates the snake evolution via Equations 2.24–2.25. Figure 2.4b shows the result of snake evolution by algorithm KWT.

2.3.1 Algorithm KWT

Step 1: Initialize snake: $\mathbf{x}^0$, $\mathbf{y}^0$

Step 2: Set: $t \leftarrow 0$

Do

- Compute $\mathbf{f}_x^\tau$ by Equation 2.14 and $\mathbf{f}_y^\tau$ by Equation 2.15
- Compute $\mathbf{x}^\tau \leftarrow (I_n + (\varsigma)A)^{-1}(\mathbf{x}^\tau + (\varsigma)\mathbf{f}_x^\tau)$
- Compute $\mathbf{y}^\tau \leftarrow (I_n + (\varsigma)A)^{-1}(\mathbf{y}^\tau + (\varsigma)\mathbf{f}_y^\tau)$
- Update counter: $t \leftarrow t + 1$

while $\left\| \mathbf{x}^{\tau+1} - \mathbf{x}^\tau \right\| + \left\| \mathbf{y}^{\tau+1} - \mathbf{y}^\tau \right\| \le \text{tolerance}$.

The snake evolution equations can also be derived completely avoiding the route of variational calculus if we make the energy functional (Equation 2.3) discrete even before minimizing it:

$$E(X_0,\ldots,X_{n-1},Y_0,\ldots,Y_{n-1}) = \frac{1}{2}\sum_{i=0}^{n}\alpha(X_{i+1} - X_i)^2 + \alpha(Y_{i+1} - Y_i)^2$$

$$+ \frac{1}{2}\sum_{i=0}^{n}\beta(X_{i+1} - 2X_i + X_{i-1})^2 + \beta(Y_{i+1} - 2Y_i + Y_{i-1})^2 - \sum_{i=0}^{n}f(X_i,Y_i). \tag{2.26}$$

Equation 2.26 is not a functional any more; it is merely a function of $2n$ variables $X_0, \ldots, X_{n-1}$, $Y_0, \ldots, Y_{n-1}$. To minimize Equation 2.26, we only need to compute partial derivatives of E with respect to the $2n$ variables and equate them to zero:

$$\frac{\partial E}{\partial X_i} = -\alpha(X_{i+1} + X_{i-1} - 2X_i)$$

$$+ \beta(X_{i+2} - 4X_{i+1} + 6X_i - 4X_{i-1} + X_{i-2}) - f_x(X_i,Y_i) \tag{2.27}$$

and

$$\frac{\partial E}{\partial Y_i} = -\alpha(Y_{i+1} + Y_{i-1} - 2Y_i) + \beta(Y_{i+2} - 4Y_{i+1} + 6Y_i - 4Y_{i-1} + Y_{i-2})$$

$$- f_y(X_i,Y_i), \forall i \in \{0,1,\ldots,n-1\}. \tag{2.28}$$

The gradient descent Equations 2.8 and 2.9 can be easily derived from Equations 2.27 and 2.28 as before.

2.4   SNAKE EXTERNAL FORCES

The active contour evolution, Equations 2.8 and 2.9, has a nice physical interpretation—that of force balance—the internal and the external forces are acting on the contour and making it move by the resultant force. In the KWT snake model, the external force is simply the force due to the potential surface created by the image gradient magnitude. With this interpretation, one naturally wonders if there are other external forces to guide snakes. In this section, we are going to examine

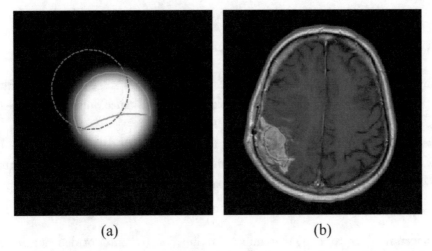

(a)   (b)

FIGURE 2.6: Illustrations of failure snakes using image gradient-based external force. Broken contour: initial snake; solid contour: final snake.

several external forces for snakes. Before jumping into their descriptions, let us see why the image gradient force used in KWT snake may not be good enough in many applications.

The limitation of the gradient external force becomes evident when we examine a homogeneous part of an image. The image gradient magnitude is essentially zero or close to zero in such portions of the image. Thus, if an active contour starts its journey here, there is little hope to guide it toward an edge (see Figure 2.6). In other words, the gradient-based force field (f_x, f_y) has a limited *capture range* for the active contour.

The distance potential force alleviates this issue for binary images. To obtain distance potential force, we first build a distance surface (or distance map) $D(x, y)$ from a binary image $I(x, y)$:

$$D(x, y) = \min_{(p,q) \in \{(a,b) : I(a,b) = 1\}} [d(x, y; p, q)], \tag{2.29}$$

where $d(x, y; p, q)$ measures the distance between two locations (x, y) and (p, q). A familiar example distance metric is Euclidean distance:

$$d(x, y; p, q) = \sqrt{(x - p)^2 + (y - q)^2}. \tag{2.30}$$

After constructing $D(x, y)$, we compute the distance potential force field $\left(-\dfrac{\partial D}{\partial x}, -\dfrac{\partial D}{\partial y}\right)$, and use this field in snake evolution as follows:

$$\frac{\partial X}{\partial \tau} = \alpha \frac{\partial^2 X}{\partial s^2} - \beta \frac{\partial^4 X}{\partial s^2} - \frac{\partial D}{\partial x}, \tag{2.31}$$

and

$$\frac{\partial Y}{\partial \tau} = \alpha \frac{\partial^2 Y}{\partial s^2} - \beta \frac{\partial^4 Y}{\partial s^2} - \frac{\partial D}{\partial y}. \tag{2.32}$$

If you compare Equations 2.31 and 2.32 with Equations 2.6 and 2.7, then you will find that the edge force (f_x, f_y) in the latter pair has been replaced by the distance potential force $(-D_x(x,y), -D_y(x,y))$.

A balloon force is an external force that tries to inflate or deflate the contour [5]. More technically speaking, the balloon force exerts either an inward or an outward force that is normal to the active contour. The normal direction to a parameterized contour at $(X(s), Y(s))$ is provided by $\left(-\frac{\partial Y}{\partial s}, \frac{\partial X}{\partial s}\right)$. The parameterization of the contour (whether s increases or decreases in a clockwise direction traversal of the contour) determines the direction (inflation or deflation) of this force. Thus, the balloon force can be denoted by $c_\tau \left(-\frac{\partial Y}{\partial s}, \frac{\partial X}{\partial s}\right)$, with a proportionality constant c_τ, which may be evolution time (τ) dependent. The snake evolution is given in the following equations:

$$\frac{\partial X}{\partial \tau} = \alpha \frac{\partial^2 X}{\partial s^2} - \beta \frac{\partial^4 X}{\partial s^4} - c_\tau \frac{\partial Y}{\partial s}, \tag{2.33}$$

and

$$\frac{\partial Y}{\partial \tau} = \alpha \frac{\partial^2 Y}{\partial s^2} - \beta \frac{\partial^4 Y}{\partial s^4} + c_\tau \frac{\partial X}{\partial s}. \tag{2.34}$$

Note that although the balloon force is an external force, it is not computed from the image. Thus, in general, it may still be a good idea to incorporate image gradient-based external force in addition to the balloon force in a snake:

$$\frac{\partial X}{\partial \tau} = \alpha \frac{\partial^2 X}{\partial s^2} - \beta \frac{\partial^4 X}{\partial s^4} + \lambda f_x(X,Y) - c_\tau \frac{\partial Y}{\partial s}, \tag{2.35}$$

and

$$\frac{\partial Y}{\partial \tau} = \alpha \frac{\partial^2 Y}{\partial s^2} - \beta \frac{\partial^4 Y}{\partial s^4} + \lambda f_y(X,Y) + c_\tau \frac{\partial X}{\partial s}, \tag{2.36}$$

where λ is an edge-force weight.

2.4.1   Gradient Vector Flow

The issue of building an adequate capture range has been very successfully addressed by the gradient vector flow (GVF) proposed by Xu and Prince [6]. GVF is an external force field $(u(x,y), v(x,y))$ constructed by diffusing the edge force (f_x, f_y). This is achieved by minimizing the following energy functional:

$$E_{\text{GVF}}(u, v) = \frac{1}{2} \iint \mu(u_x^2 + u_y^2 + v_x^2 + v_y^2) + (f_x^2 + f_y^2)$$

$$((u - f_x)^2 + (v - f_y)^2)\mathrm{d}x\mathrm{d}y,$$

(2.37)

where μ is a non-negative parameter controlling the degree of smoothness of the field (u,v). Notice that the first integrand keeps the field, (u,v), smooth, while the second integrand encourages the field to resemble the edge force where the latter is strong. The following two Euler equations are obtained after applying variational minimization to Equation 2.37:

$$\mu\nabla^2 u - (f_x^2 + f_y^2)(u - f_x) = 0,$$

(2.38)

and

$$\mu\nabla^2 v - (f_x^2 + f_y^2)(v - f_y) = 0.$$

(2.39)

Figure 2.7a shows the GVF force computed on the circle image of Figure 2.1a by solving the partial differential Equations 2.38–2.39. Notice that now, even within homogeneous zones, the external force is present. How do we use the force field in snake evolution? Simply replacing the external forces by (u,v) in the snake evolution equations:

$$\frac{\partial X}{\partial \tau} = \alpha \frac{\partial^2 X}{\partial s^2} - \beta \frac{\partial^4 X}{\partial s^2} + u(X, Y),$$

(2.40)

and

$$\frac{\partial Y}{\partial \tau} = \alpha \frac{\partial^2 Y}{\partial s^2} - \beta \frac{\partial^4 Y}{\partial s^2} + v(X, Y).$$

(2.41)

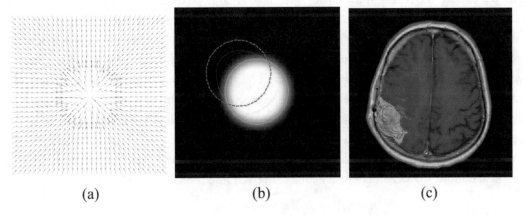

(a) (b) (c)

FIGURE 2.7: (a) GVF force field on the circle image of Figure 2.1a. (b) Snake evolution via GVF. Same initial contour as in Figure 2.6a is used here. (c) GVF snake evolution on a brain MR image. Same initial contour as in Figure 2.6b is used.

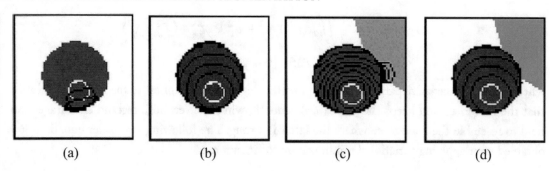

(a)   (b)   (c)   (d)

FIGURE 2.8: (a) Initial snake (white contour) and evolution by GVF (black contours). (b) Initial snake (white contour) and EGVF snake evolution (black contours). (c) Initial snake (white contour) and snake evolution by balloon force (black contours). (d) Initial snake (white contour) and EGVF snake evolution (black contours). Taken from Reference [7].

Figure 2.7b shows snake evolution by GVF force field. The same initial contour is used here. Unlike the image gradient force, the snake captures the object. A similar result is observed for the magnetic resonance imaging in Figure 2.7c. These examples illustrate that GVF can attract even a faraway snake toward object edges.

While GVF effectively increases the "capture range" for the snake, we still cannot guarantee correct object segmentation from any initial snake. We illustrate this point by means of a toy example in Figure 2.8a. Notice here that the snake fails to delineate the object boundary. Looking at the GVF fields (see Figure 2.7a) immediately reveals the reason behind its failure. The flow vectors seem to emanate from the center of the circle, in this case. So, it is obvious that if the initial snake does not enclose the center of the circle, then there is no hope of the snake to delineate the object boundary. One remedy here might be that if we can somehow ensure the placement of the initial snake inside the object, then an inflating balloon force can eventually delineate the object boundary. However, the balloon force can cause considerable trouble (see Figure 2.8c). With balloon force, the snake "leaks" through edges having weak gradient magnitude. Actually, this is not surprising, since the balloon force inflates the snake everywhere with an equal amount of force completely disregarding the image data, i.e., the gradient strength. These examples force us to think whether we can create an inflating snake that is not as blind as the balloon snake. Thus, the question is: Can we inflate the snake in a way that it sees or senses the image data ahead of it? Indeed, we can do it. The trick here is to impose a Dirichlet boundary condition (BC) into the GVF PDEs [7]:

$$\mu \nabla^2 u - (f_x^2 + f_y^2)(u - f_x) = 0,$$
$$\mu \nabla^2 v - (f_x^2 + f_y^2)(v - f_y) = 0,$$
$$\text{when } (x, y) \in (\Omega \backslash C), (u, v) = \lambda \mathbf{n}_{\partial C}$$
$$\text{when } (x, y) \in \partial C, \ \nabla u.\mathbf{n}_{\partial C} = 0, \text{ and } \nabla v.\mathbf{n}_{\partial C} = 0, \tag{2.42}$$

where Ω denotes the image domain, C denotes the circular domain enclosed by the initial active contour (assumed to be a circle), $\Omega \setminus C$ denotes the set difference of Ω and C, $\partial\Omega$ and ∂C are, respectively, the boundaries of Ω and C, and $\mathbf{n}_{\partial C}$ and $\mathbf{n}_{\partial\Omega}$ are unit outward normal to the boundaries ∂C and $\partial\Omega$, respectively, μ is a non-negative parameter controlling the smoothness of the vector field, λ is a positive parameter (Dirichlet BC), and f is the edge-map for the image, such as $|\nabla I|^2$. We refer to the solution (u,v) of the PDEs (Equation 2.42) as *enhanced* gradient vector flow (EGVF) field.

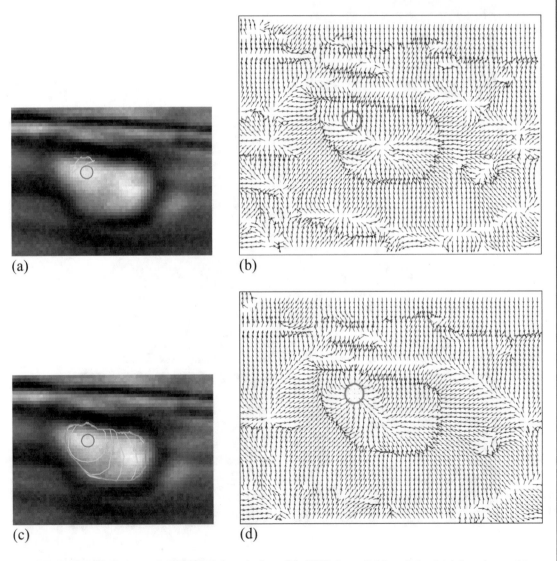

(a)

(b)

(c)

(d)

FIGURE 2.9: (a) Failure of GVF snake evolution. (b) GVF force field and the initial snake position. (c) EGVF snake evolution. (d) EGVF force field and the initial snake.

Figure 2.8b illustrates the action of an EGVF snake on the circle image with initial snake shown in Figure 2.8a. Similarly, in Figure 2.8d, the EGVF snake is able to segment the object, where the balloon snake falters. Moving to a more realistic example, in Figure 2.9, we show a leukocyte

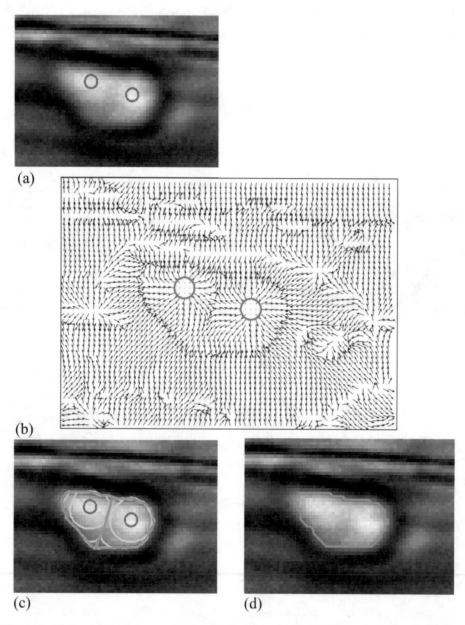

FIGURE 2.10: (a) Leukocyte image and two initial snakes. (b) EGVF force field. (c) EGVF snake evolution: red—initial, yellow—intermediate, and green—final contours are shown. (d) Final segmentation after merging the evolved snakes.

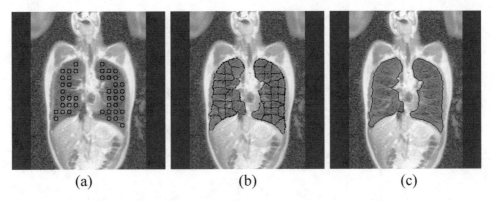

  (a)   (b)   (c)

FIGURE 2.11: (a) Lung MR image and a number of initial snakes inside the lung cavity. (b) Fully evolved EGVF snakes. (c) Merging the evolved snakes produce lung segmentation.

boundary segmentation application. Visual comparison of GVF field (Figure 2.9a) and EGVF field (Figure 2.9b) reveals why EGVF snake works here, while the GVF snake fails.

      Another interesting property of EGVF snake evolution is automated merging of parametric snakes. Let us image that instead of a single parametric active contour, we have managed to initialize two contours as shown in Figure 2.10a. Then, we solve Equation 2.42. Note that there are two separate boundaries where the Dirichlet condition must be imposed in Equation 2.52. After solving for the EGVF field (shown in Figure 2.10b), we evolve two snakes *independently* to obtain the evolved contours (see Figure 2.10c). Next, by taking the set union of the two areas bounded by the two fully evolved snakes, we obtain the leukocyte boundary shown in Figure 2.10d. Figure 2.11 illustrates a lung segmentation application with EGVF snakes.

2.4.2   Vector Field Convolution Active Contours

The idea behind gradient vector flow [6] is quite appealing—to diffuse the active contour external forces away from the object so that the initialization is more forgiving. We extend this idea by computing the diffused external force via convolution. Vector field convolution (VFC) uses a prefixed vector kernel that is convolved with the edge map to make an extended external force in the form of a vector field. VFC defines a novel external force field with a large capture range and the ability to converge to concavities. As compared to gradient vector flow [6], VFC yields improved robustness to noise and initialization, a more flexible force field definition, and reduced computational cost [8].

      The prefixed vector kernel **k** is defined as

$$\mathbf{k}(x,y) = m(x,y)\,\mathbf{n}(x,y), \qquad\qquad (2.43)$$

where $m(x,y)$ is the magnitude of the vector at (x,y), and $\mathbf{n}(x,y)$ is the unit vector pointing to the origin (note that the origin of the discrete kernel is located at the center of the matrix **k**). An ex-

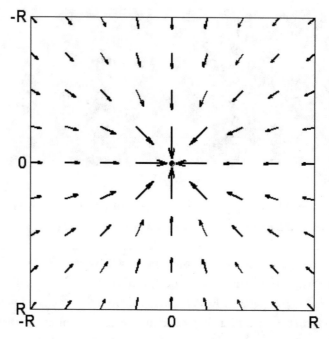

FIGURE 2.12: An example of the VFC kernel $\mathbf{k}(x,y)$ for a discrete down with a maximum radius R of 4.

ample $\mathbf{k}(x,y)$ is shown in Figure 2.12. Instead of a diffusion process, the VFC external force field $\mathbf{v}(x,y)$ is yield by the convolution of the vector field kernel $\mathbf{k}(x,y)$ and the edge map $f(x,y)$. The edge map can be as simple as the map derived by computing a normalized gradient magnitude, or it can be more sophisticated for specialized applications. If we split $\mathbf{k}$ into its x-component $k_x(x,y)$ and its y-component $k_y(x,y)$, the convolution of VFC is given by:

$$v(x,y) = f(x,y) * \mathbf{k}(x,y) = [\, f(x,y) * k_x(x,y), f(x,y) * k_y(x,y)]. \qquad (2.44)$$

As the edge map $f(x,y)$ has larger values near the image edges, these edges contribute more to the VFC force field than do the homogeneous regions. Further, since the objects have sustained edges, they contribute more to the force field than do clutter and noise.

The fact that the kernel $\mathbf{k}(x,y)$ should be isotropic, in the sense that no direction is favored, and each vector emanates from the center, is quite obvious. However, the choice of VFC kernel vector magnitude $m(x,y)$ is not so obvious. The VFC field is highly dependent on the choice of the vector field kernel magnitude $m(x,y)$. By considering the fact that the influence from the feature of interest should be less as the particles are further away, the magnitude should be a decreasing function of distance from the origin:

$$m(x,y) = (r + \varepsilon)^{-\gamma}, \qquad (2.45)$$

where γ is a positive parameter that controls the rate of decrease in magnitude. The radius r from the center of the kernel is equal to $\sqrt{x^2 + y^2}$. Here, ε is a small positive constant that prevents division by zero at the origin. This formulation is inspired by Newton's law of universal gravitation in physics which can be viewed as a special case of Equation 2.45 with $\gamma = 2$ and $\varepsilon = 0$. It should be noted that, although limited research has been performed on selecting $m(x,y)$ in the VFC operation, more research is merited for the impact on particular applications.

So, as illustrated for the MR image of an ankle given in Figure 2.13a, the VFC process begins with an edge map (Figure 2.13b) and leads to an extended force field as shown in Figure 2.13c. The lines shown in Figure 2.13c are *streamlines*. One can imagine a single active contour element (a "snaxel") at a given point in space. The streamline shows the path of that singular element before it ceases movement. So, given an initial contour (as shown in Figure 2.13d), the VFC external force field can guide the contour to the boundary, which, in this case, is the cartilage–bone boundary of the ankle (see Figure 2.13e).

2.4.3   Inverse Problem Approach to Active Contour Initialization

According to the inverse problem theoretician Alifanov, "Solution of an inverse problem entails determining unknown causes based on observation of their effects." So, in the case of active contour initialization, the *effects* are the external force vectors that are *caused* by the object boundary. We would like to approximate this object boundary given the external force vectors. This is an inverse problem view of active contour initialization.

There are three fundamental options for initializing an active contour: (1) naïve initialization, (2) manual initialization, and (3) automatic initialization. Naïve initialization defines the initial contour without knowledge of image content, e.g., one may initialize the snake at the image boundary or as a simple geometric shape. Such initializations are generally distant from the object of interest and require an inordinate number of iterations to capture the desired boundary. Further, such a naïve initial contour may lead to divergence, in which the incorrect object is captured, or clutter "distracts" the snake from the feature of interest. With manual initialization, the user selects a coarse object boundary (or an initial interior point), which is then refined by an active contour algorithm. Such manual interaction is tedious and time-consuming, and in applications such as high-content screening (HCS), manual initialization is not a feasible option. Furthermore, in other applications that involve three-dimensional imagery, manual initialization is difficult if not impossible.

The third option available is, of course, automatic active contour initialization. In addition to the method developed in our laboratory, there are at least two automated initialization options for active contours and surfaces. The first method, termed *centers of divergence* (CoD) [9], places small circular initial contours at points of zero vector divergence within a given external force field, such as those provided by gradient vector flow [6] or VFC [8]. Like the inverse problem approach

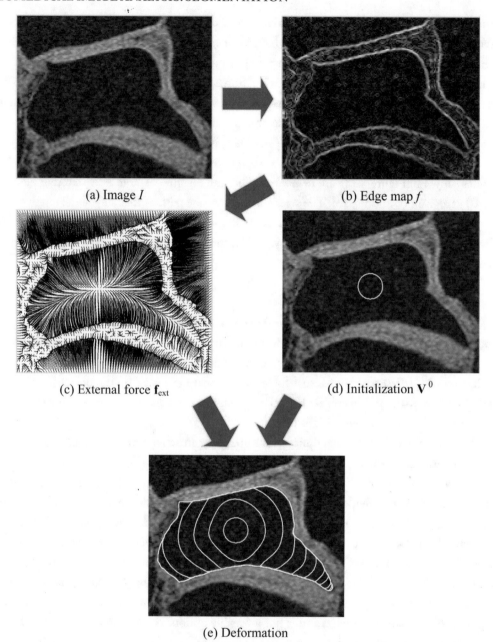

(a) Image I

(b) Edge map f

(c) External force $\mathbf{f}_{\text{ext}}$

(d) Initialization $\mathbf{V}^0$

(e) Deformation

FIGURE 2.13: (a) MR image of human ankle; (b) the edge map given by gradient magnitude; (c) the external force (shown by streamlines) computed from the edge map; (d) an initialization; (e) active contour deformation using VFC to capture the object boundaries.

presented here, the CoD method is well grounded in using the external force to form an initial segmentation, as the vectors within the force field are pointing toward objects of interest. However, the CoD suffers from oversegmentation and often requires significant postprocessing in the form of region merging.

The other automated method, termed *force field segmentation* (FFS) [10], quantizes the external force vector field (into say, eight directional unit vectors) and forms connected components in between opposing vectors to define the initialization for a system of snakes. This approach, while also exploiting the external force system, is sensitive to clutter and broken edges and generally leads to spurious contours.

Our approach, termed Poisson inverse gradient (PIG) initialization, is essentially the generation of a coarse segmentation that has an associated energy which is close to the global minimum [11]. Rather than using the traditional segmentation framework that first defines an external force field and subsequently evolves an active contour to delineate the object boundary, the PIG method instead attempts to solve the inverse problem of determining the object boundary that produced a given external force field. As mentioned, we are estimating the boundary that caused the given external force vectors.

Critical to this technique is the calculation of an external energy field $E(x,y)$ associated with a given force field $\mathbf{f}(x,y)$. This energy field has the same domain as the image and maps the locations of the image plane to a scalar energy. We find lines of constant value, termed *isolines*, within this energy field $\mathbf{E}$ and define the minimum isoline (isoline of minimum energy level) as our active contour initialization. This minimal isoline represents a minimal energy approximation of the true optimal contour.

As may be derived via the Euler–Lagrange equations, the relationship between the external force and the external energy is given by

$$\mathbf{f} = -\nabla E. \tag{2.46}$$

Therefore, we seek the energy $E(x,y)$ that has a negative gradient that is close to $\mathbf{f}(x,y)$ in the L_2 sense:

$$E(x,y) = \arg\min_{E} \iint \left| -\nabla E(x,y) - \mathbf{f}(x,y) \right|^2 dx\,dy. \tag{2.47}$$

In electromagnetic theory, this Equation 2.47 can be used to find the electric potential given a particular electric charge distribution. In image processing, we can use the solution to find an initial segmentation. This solution, naturally, is given by Poisson's equation:

$$\Delta E(x,y) = -\mathrm{div}\, f(x,y). \tag{2.48}$$

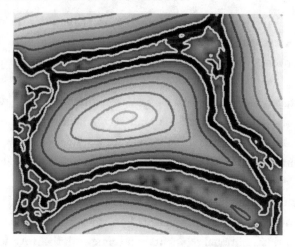

FIGURE 2.14: For the MR image of Figure 2a, this figure shows the isolines of the external energy estimated using PIG.

On a discrete grid, we may use any of several methods developed for banded sparse matrices to obtain E in Equation 2.48. To mention a few, solutions include successive over-relaxation [12], conjugate gradient techniques [13], fast Fourier transform (FFT) techniques [14], and multigrid methods [15].

Once the energy is computed using Equation 2.48, the initial contour in two dimension is computed by finding a close isoline of minimum energy. Such isolines are shown for an MR image of the ankle in Figure 2.14. For three-dimensional operation, isosurfaces can be utilized. Note that isolines or isosurfaces may be excluded by way of certain criteria such as size and shape.

We have validated the inverse initialization approach in synthetic image tests [11]. For each test, the CoD method, the FFS method, and the PIG method were compared. The inverse method (PIG) excels over the other two automated initializations in fidelity for objects with broken edges, where the FFS method has particular difficulties. In terms of objects of increasing curvature, it is the FFS method that encounters the greatest error, and once again, the inverse method presented here is superior. Finally, the PIG method improves upon both the RMSE and the computational expense of computing an initial segmentation in noise [11].

2.4.4 Feature-Weighted Snakes

Typically, the capture range of a snake or active contour/surface is extended by means of smoothing or diffusing the forces arising from the gradient. Vector field convolution [8] and gradient vector flow [6] are examples of this strategy. In this section, we take an alterative strategy. Here, we boost the "attraction" of the snake to the desired boundary by seeking to capture an object of a given area or shape. This approach results in *feature-weighted (FW) snakes* [16].

The FW active contour external energy is a pointwise product of the original external energy and a feature-based weighting. This new external energy is expressed by

$$E_{FW}(x, y) = w(S(x, y)) E_{ext}(x, y) \qquad (2.49)$$

where $E_{ext}(x, y)$ is the (nonpositive) standard external energy (e.g., $E_{ext}(x, y) = -|\nabla I(x, y)|$), and $w(S(x, y))$ is the *weighting function* of the *feature score* $S(x, y)$ at position (x, y). This feature score can be used to emphasize objects of a certain area or shape. The score is then weighted by w, which is bound below by 0 and above by 1.

2.4.5   Area-Weighted Snakes

A salient observation is made by noting that noise and small clutter do not have substantial interior area. Therefore, we can make the force resultant from the external energy grow with area such that the snake is not attracted to noise and clutter. An *area-based* (AB) weighting function can be constructed using

$$w_{AB}(S(x, y)) = 1 - \exp\left\{-S(x, y)^2 \big/ k^2\right\} \qquad (2.50)$$

where $S(x, y)$ is the feature score that is commensurate with the number of connected edge pixels. Here, k serves as a soft threshold that determines a significant size in terms of object perimeter. Thus, Equation 2.50 enacts a soft selection of objects instead of an all-out appearance/disappearance function as with area morphology, for example.

We observe that the AB weighting function w_{AB} is close to one if the area of the connected strong features is much larger than k and close to zero if the area is small. An example of the area-based weighting function is shown in Figure 2.15.

2.4.6   Correlation-Weighted Snakes

When seeking to segment an object of known shape, the constituent edges may be broken or weak, or the snake could be attracted by the edges from clutter. By way of the prior shape model, we can avoid the ill effects of clutter, noise, or broken edges.

Given an M-pixel binary template T, we can compute a cross-correlation-based (CCB) score and a weighting based on that score. The score is computed using

$$S = \frac{1}{M}(-E_{ext} \otimes T) \oplus T^{\dagger} \qquad (2.51)$$

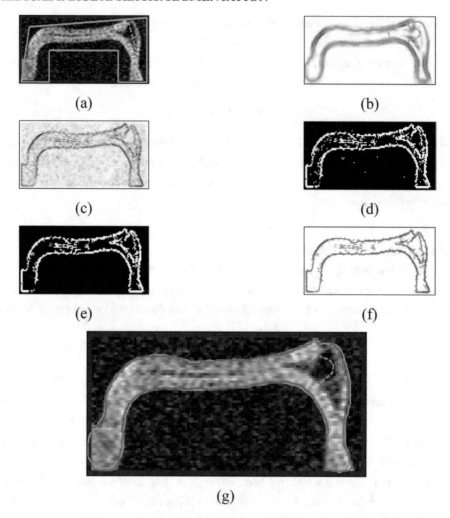

FIGURE 2.15: (a) A magnetic resonance image of the human ankle cartilage and two independent initial snakes; (b) the standard (gradient-based) external energy and snake results; (c) the associated external energy; (d) the strong features (white); (e) the area-based weighting function w_{AB} ; (f) the area-based external energy and snake results; (g) the image superposed with snake results using the standard external energy (green) and the area-based external energy (red) [16].

where $\otimes$ denotes cross-correlation operation, and $\oplus$ denotes morphological dilation. $T^{\dagger}$ represents the 180-degree rotation of T.

The CCB weighting function is then computed as

$$w_{\mathrm{CCB}}\left[S(x,y)\right] = \begin{cases} 0 & S(x,y) < c \\ \dfrac{S(x,y) - c}{\eta} & S(x,y) \geq c \end{cases} \tag{2.52}$$

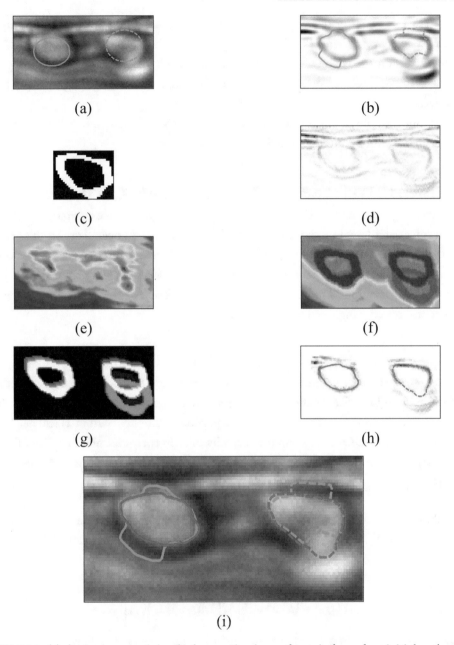

FIGURE 2.16: (a) An image containing leukocytes in vivo and two independent initial snakes; (b) the standard (gradient-based) external energy and snake results; (c) the edge template used for the CCB weighting function; (d) the associated external energy; (e) the cross-correlation result and (f) the CCB feature score S (red and blue represent high and low values, respectively); (g) the CCB weighting function w_{CCB} ; (h) the CCB external energy and snake results; (i) the image superposed with snake results using the standard external energy (green) and the CCB external energy (red) [16].

where $n = |T| - c$ is a normalization constant, c is the minimum desired feature score, and $|T|$ is the cardinality of nonzero elements in T. If an edge matches the edge template T, the feature scores S at all the pixels on the edge will be relatively high. In this case, the CCB weighting function w_{CCB} is close to one at all the pixels on the edge. The range of the weighting function is zero to one inclusive.

An example of the cross-correlation-based snake is shown in Figure 2.16. Here, two leukocytes are located within a cluttered intravital microscopy image.

2.5   SNAKES WITH SPECIAL PARAMETERIZATION

The last section lays out an important component in active contour/snake computing—the concept of external forces and how to build them to make best use of the image data. In this section, we embark on the other counterpart of the snake—the internal energy. Internal energy of a contour is essentially another name for the *flexibility* of a snake. The flexibility of a contour depends on how we parameterize the contour. The goal of this section is to discuss a few different parameterizations for active contours and tying them with different applications.

2.5.1   Spline Snakes: Open, Closed, and Clamped B-Splines

In a general sense, any parameterized curve can be termed as a snake. The flexibility of the curve is characterized by a local deformation behavior—if a point on the curve is given a small displacement, then the entire curve is not perturbed; only a portion of the curve within a local neighborhood of the point is deformed. One very important family of such curves is cubic B-splines that are C^2 continuous curves [17]. A curve $(X(s), Y(s))$ parameterized via s is called C^n continuous when both $X(s)$ and $Y(s)$ are continuous, and the derivatives $\frac{dX}{ds}, \frac{d^2X}{ds^2} \cdots, \frac{d^nX}{ds^n}$ and $\frac{dY}{ds}, \frac{d^2Y}{ds^2} \cdots, \frac{d^nY}{ds^n}$ are also all continuous. A cubic B-spline $(X(s), Y(s))$ where $s \in [0,1]$ can be defined via a sequence of $(N+3)$ "control" points $(X_1, Y_1), (X_2, Y_2), \ldots, (X_{N+3}, Y_{N+3})$, with integer $N > 0$ as follows:

$$X(s_{i,u}) = [u^3 \ u^2 \ u \ 1]M[X_i \ X_{i+1} \ X_{i+2} \ X_{i+3}]^T,$$

$$Y(s_{i,u}) = [u^3 \ u^2 \ u \ 1]M[Y_i \ Y_{i+1} \ Y_{i+2} \ Y_{i+3}]^T,$$

(2.53)

where the scalar parameter $u \in [0, 1]$ and the coefficient matrix M is as follows:

$$M = \frac{1}{6}\begin{bmatrix} -1 & 3 & -3 & 1 \\ 3 & -6 & 3 & 0 \\ -3 & 0 & 3 & 0 \\ 1 & 4 & 1 & 0 \end{bmatrix}.$$

(2.54)

The parameter $s_{i,u}$ in Equation 2.53 is defined as:

$$s_{i,u} = \frac{i - 1 + u}{N}, \tag{2.55}$$

and range of i is $1 \leq i \leq N$. Note that $s_{i,u} \in \left[\dfrac{i-1}{N}, \dfrac{i}{N}\right]$ and $\bigcup\limits_{i=1}^{N} \left[\dfrac{i-1}{N}, \dfrac{i}{N}\right] = [0,1]$; so the entire

curve with $s \in [0,1]$ is divided into N segments and the ith segment is defined by Equations 2.53. Further, it can be easily verified that the end point of the ith segment of the curve is the starting point of the $(i+1)$th segment of the curve. Starting and end points of B-spline segments are called knot points. The local deformation behavior of a cubic B-spline is now evident. Since four control points $(X_i, Y_i), (X_{i+1}, Y_{i+1}), (X_{i+2}, Y_{i+2})$, and (X_{i+3}, Y_{i+3}) define the ith segment, or equivalently (X_i, Y_i) appears in the ith, $(i-1)$th, $(i-2)$th, and $(i-3)$th segments of the B-spline, displacing one control point (X_i, Y_i) will only affect, at the most, these four curve segments.

A cubic B-spline curve can be closed. One way to obtain a closed B-spline curve is as follows: make the first three control point the same as the last three control points, i.e., when $X_1 = X_{N+1}$, $X_2 = X_{N+2}, X_3 = X_{N+3}$ and $Y_1 = Y_{N+1}, Y_2 = Y_{N+2}, Y_3 = Y_{N+3}$, the spline is closed. Several other variations of open and closed B-splines can be obtained by suitably specifying the control points. One such useful variation is obtained by clamping a knot point. For example, if we want to clamp the first knot point $(X(s_{1,0}), Y(s_{1,0}))$ to the control point (X_1, Y_1), then a "phantom" control point (X_0, Y_0) has to be introduced for the first B-spline segment:

$$\begin{aligned}
X(s_{1,u}) &= [\, u^3 \ u^2 \ u \ 1\,] M [\, X_0 \ X_1 \ X_2 \ X_3\,]^\mathrm{T}, \\
Y(s_{1,u}) &= [\, u^3 \ u^2 \ u \ 1\,] M [\, Y_0 \ Y_1 \ Y_2 \ Y_3\,]^\mathrm{T},
\end{aligned} \tag{2.56}$$

where (X_0, Y_0) can be found by:

$$\begin{aligned}
X_1 &= X(s_{1,0}) = [\,0\,0\,0\,1\,] M [\, X_0 \ X_1 \ X_2 \ X_3\,]^\mathrm{T} \Rightarrow X_0 = 2X_1 - X_2, \\
Y_1 &= Y(s_{1,0}) = [\,0\,0\,0\,1\,] M [\, Y_0 \ Y_1 \ Y_2 \ Y_3\,]^\mathrm{T} \Rightarrow Y_0 = 2Y_1 - Y_2.
\end{aligned}$$

Therefore, the first segment of this B-spline has the form:

$$\begin{aligned}
X(s_{1,u}) &= [\, u^3 \ u^2 \ u \ 1\,] M [(2X_1 - X_2) \ X_1 \ X_2 \ X_3]^\mathrm{T}, \\
Y(s_{1,u}) &= [\, u^3 \ u^2 \ u \ 1\,] M [(2Y_1 - Y_2) \ Y_1 \ Y_2 \ Y_3]^\mathrm{T},
\end{aligned} \tag{2.57}$$

and the remaining $(N-1)$ segments of the B-spline have the following form:

$$\begin{aligned}
X(s_{i,u}) &= [\, u^3 \ u^2 \ u \ 1] M [X_{i-1} \ X_i \ X_{i+1} \ X_{i+2}]^\mathrm{T}, \\
Y(s_{i,u}) &= [\, u^3 \ u^2 \ u \ 1] M [Y_{i-1} \ Y_i \ Y_{i+1} \ Y_{i+2}]^\mathrm{T}, \ 2 \leq i \leq N.
\end{aligned} \tag{2.58}$$

The sequence of control points for the clamped-knot B-spline Equations 2.57–2.58 is $(2X_1 - X_2, 2Y_1 - Y_2), (X_1, Y_1), (X_2, Y_2), \ldots, (X_{N+2}, Y_{N+2})$. A similar trick can clamp the last knot point of the B-spline to (X_1, Y_1) as well. In that case, we have a closed B-spline with a clamped-knot point.

The control point sequence for a closed B-spline with one clamped-knot point is $(2X_1 - X_2, 2Y_1 - Y_2)$, (X_1, Y_1), (X_2, Y_2), ... , (X_N, Y_N), (X_1, Y_1), $(2X_1 - X_N, 2Y_1 - Y_N)$. These B-spline parameterization models play significant roles segmenting objects, as we illustrate shortly. Many other interesting cases are obtained by manipulating the control point sequence. Interested readers are referred to Reference [17].

2.5.2   Gradient Descent Method for Spline Snake Computation

Having defined the cubic B-spline representation, we now turn our attention to utilize this family of curves as snakes for object delineation. Again, the required tool here is the energy minimization. Let $(X(s), Y(s))$ with $s \in [0,1]$ represent a cubic B-spline. The B-spline snake energy functional is then written as:

$$E_{sp} = -\int_0^1 f(X(s), Y(s))\, ds, \qquad (2.59)$$

where f is typically an edge indicator function such as $f(x,y) = |\nabla I(x,y)|^2$. Notice the difference between Equation 2.59 and the snake energy functional Equation 2.3 is the absence of internal energy term in the former. The internal energy is implicit in the spline representation because the spline form imposes C^2 type smoothness on the curve. If the spline is represented via N segments, then Equation 2.59 can be expressed as:

$$E_{sp} = -\sum_{i=1}^{N} \int_{\frac{i-1}{N}}^{\frac{i}{N}} f(X(s_{i,u}), Y(s_{i,u}))\, ds_{i,u}. \qquad (2.60)$$

Further change of variable from $s_{i,u}$ to u turns Equation 2.60 into:

$$E_{sp} = -\frac{1}{N}\sum_{i=1}^{N} \int_0^1 f\left(X\left(\frac{i-1+u}{N}\right), Y\left(\frac{i-1+u}{N}\right)\right) du. \qquad (2.61)$$

E_{sp} is a function (not a functional!) of the control points. So we seek to minimize Equation 2.61 by varying the position of the control points via gradient descent method. The update rule for control point positions via gradient descent equations are:

$$X_i^{\tau+1} = X_i^{\tau} - \delta\tau \frac{\partial E_{sp}}{\partial X_i},$$

$$Y_i^{\tau+1} = Y_i^{\tau} - \delta\tau \frac{\partial E_{sp}}{\partial Y_i}, \qquad (2.62)$$

where the superscripts τ and $\tau+1$ are two consecutive iterations of the gradient descent update, and $\delta\tau$ is a suitably chosen time step. The energy minimization is an iterative process that repeatedly computes control points via Equation 2.62, until there are any appreciable changes in them. The trickiest part here is choosing the step size $\delta\tau$ that typically involves some trial and error tasks.

Some details are as follows when we compute the partial derivatives $\dfrac{\partial E_{sp}}{\partial X_i}$ and $\dfrac{\partial E_{sp}}{\partial Y_i}$. To do so, we first write from Equation 2.61:

$$
\begin{aligned}
\frac{\partial E_{sp}}{\partial X_i} &= -\frac{1}{N}\frac{\partial}{\partial X_i}\left[\sum_{k=i-3}^{i}\int_0^1 f\left(X\left(\frac{k-1+u}{N}\right), Y\left(\frac{k-1+u}{N}\right)\right)du\right]\\
&= -\frac{1}{N}\sum_{k=i-3}^{i}\int_0^1 \frac{\partial}{\partial X_i}\left[f\left(X\left(\frac{k-1+u}{N}\right), Y\left(\frac{k-1+u}{N}\right)\right)\right]du \qquad (2.63)\\
&= -\frac{1}{N}\sum_{k=i-3}^{i}\int_0^1 \frac{\partial f}{\partial x}\left(X\left(\frac{k-1+u}{N}\right), Y\left(\frac{k-1+u}{N}\right)\right)\frac{\partial}{\partial X_i}\left[X\left(\frac{k-1+u}{N}\right)\right]du.
\end{aligned}
$$

Now, with the help of Equation 2.53, writing the derivatives $\dfrac{\partial}{\partial X_i}X\left(\dfrac{k-1+u}{N}\right)$ explicitly in Equation 2.63, we obtain:

$$
\begin{aligned}
\frac{\partial E_{sp}}{\partial X_i} = -\frac{1}{6N}\Bigg[&\int_0^1 \frac{\partial f}{\partial x}\left(X\left(\frac{i-4+u}{N}\right), Y\left(\frac{i-4+u}{N}\right)\right)u^3\,du\\
&+\int_0^1 \frac{\partial f}{\partial x}\left(X\left(\frac{i-3+u}{N}\right), Y\left(\frac{i-3+u}{N}\right)\right)(-3u^3+3u^2+3u+1)du\\
&+\int_0^1 \frac{\partial f}{\partial x}\left(X\left(\frac{i-2+u}{N}\right), Y\left(\frac{i-2+u}{N}\right)\right)(3u^3-6u^2+4)du\\
&+\int_0^1 \frac{\partial f}{\partial x}\left(X\left(\frac{i-1+u}{N}\right), Y\left(\frac{i-1+u}{N}\right)\right)(-u^3+3u^2-3u+1)du\Bigg].
\end{aligned}
$$

(2.64)

A similar expression exists for $\dfrac{\partial E_{sp}}{\partial Y_i}$:

$$
\begin{aligned}
\frac{\partial E_{sp}}{\partial Y_i} = -\frac{1}{6N}\Bigg[&\int_0^1 \frac{\partial f}{\partial y}\left(X\left(\frac{i-4+u}{N}\right), Y\left(\frac{i-4+u}{N}\right)\right)u^3\,du\\
&+\int_0^1 \frac{\partial f}{\partial y}\left(X\left(\frac{i-3+u}{N}\right), Y\left(\frac{i-3+u}{N}\right)\right)(-3u^3+3u^2+3u+1)du
\end{aligned}
$$

$$+ \int_0^1 \frac{\partial f}{\partial y}\left(X\left(\frac{i-2+u}{N}\right), Y\left(\frac{i-2+u}{N}\right)\right)(3u^3 - 6u^2 + 4)du$$

$$+ \int_0^1 \frac{\partial f}{\partial y}\left(X\left(\frac{i-1+u}{N}\right), Y\left(\frac{i-1+u}{N}\right)\right)(-u^3 + 3u^2 - 3u + 1)du \Bigg].$$

(2.65)

We utilize these treatments of spline snakes next in a case study.

2.5.3   Segmenting Leukocytes via Spline Snakes

One important task in snake segmentation is the placement of the initial contour starting from which the iterative energy minimization process proceeds. In this section, we illustrate an application—extraction of rolling leukocyte shape from intravital microscopy video. In connection with this application, we illustrate initialization for the spline snake and subsequent object delineation. The rolling leukocytes are seen to be somewhat teardrop-shaped. To initialize a contour, we need two tools—a teardrop-shaped template and a score associated with the template computed from image data. The score should ideally have a discriminating power, so that it is able to say when the template has a maximum overlap with a leukocyte on the image. One such teardrop-shaped parametric shape can de defined via the following polar coordinate equations (r, θ) [18]:

$$r(\theta; A, n, \omega) = \frac{A}{\{|\sin((\theta + \omega)/4)|^n + |\cos((\theta + \omega)/4)|^n\}^{1/n}}, \quad 0 \leq \theta \leq 2\pi, \quad (2.66)$$

where A is a parameter controlling the size of the teardrop, n is another parameter controlling the position of cusp (trailing edge) in the teardrop shape. Both A and n are positive real numbers. ω (in radians) specifies the orientation of the teardrop. Figure 2.17 shows two such teardrop shapes

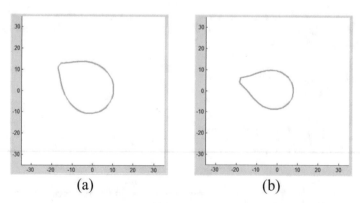

(a)   (b)

FIGURE 2.17: Teardrop models with different parameters. (a) $A = 23$, $N = 0.6$, $\omega = -45°$. (b) $A = 30$, $N = 0.45$, $\omega = -15°$. Taken from Reference [20].

for different values of A, n, and ω. A special case of the teardrop model (Equation 2.66) is a circle, which is obtained for $n = 2$. If we want the teardrop to be displaced by (t_x, t_y), then we need to convert the polar coordinates to Cartesian and add (t_x, t_y) to the Cartesian coordinates:

$$X(\theta; A, n, \omega) = r(\theta; A, n, \omega)\cos(\theta) + t_x,$$

$$Y(\theta; A, n, \omega) = r(\theta; A, n, \omega)\sin(\theta) + t_y. \qquad (2.67)$$

We can combine Equations 2.66 and 2.67 to write explicitly the teardrop in Cartesian coordinates with five parameters (A, n, ω, t_x, t_y) as follows:

$$X(\theta; A, n, \omega, t_x) = \frac{A\cos(\theta)}{\{|\sin((\theta + \omega)/4)|^n + |\cos((\theta + \omega)/4)|^n\}^{1/n}} + t_x,$$

$$Y(\theta; A, n, \omega, t_y) = \frac{A\sin(\theta)}{\{|\sin((\theta + \omega)/4)|^n + |\cos((\theta + \omega)/4)|^n\}^{1/n}} + t_y, \quad 0 \le \theta \le 2\pi. \qquad (2.68)$$

Having defined the teardrop-shaped template, our next task is to define a score that helps to determine for what values of the aforementioned five parameter values the template has maximum overlap with the leukocyte. A highly discriminating scoring statistic is known as gradient inverse coefficient of variation (GICOV) [19]. For an image $I(x,y)$ and for a curve $(X(s), Y(s))$ with $s \in [0, 1]$, GICOV is defined as:

$$g = \frac{\mu}{\sigma}, \qquad (2.69)$$

where μ and σ are, respectively, the mean and the standard deviation of the image directional derivatives along the curve $(X(s), Y(s))$. The directional derivatives are taken in the outward normal direction of the curve. Thus, μ and σ are defined as follows:

$$\mu = \int_0^1 \nabla I(X(s), Y(s)) \cdot \mathbf{n}(X(s), Y(s))\,ds,$$

$$\sigma^2 = \int_0^1 [\nabla I(X(s), Y(s)) \cdot \mathbf{n}(X(s), Y(s))]^2 ds - \mu^2, \qquad (2.70)$$

where $\mathbf{n}(X(s), Y(s)) \equiv (n_x X(s), Y(s)), n_y X(s), Y(s)))$ is the unit outward normal to the curve at $(X(s), Y(s))$ and is defined as:

$$n_x(X(s), Y(s)) = \frac{dY}{ds} \Big/ \sqrt{\left(\frac{dX}{ds}\right)^2 + \left(\frac{dX}{ds}\right)^2},$$

$$n_y(X(s), Y(s)) = -\frac{dX}{ds} \Big/ \sqrt{\left(\frac{dX}{ds}\right)^2 + \left(\frac{dX}{ds}\right)^2}, \qquad (2.71)$$

where, we can write $\dfrac{dX}{ds}$ and $\dfrac{dX}{ds}$ as follows (assuming the curve to be a B-spline):

$$\frac{\mathrm{d}X}{\mathrm{d}s} = N[3u^2\ 2u\ 1\ 0]M[X_i\ X_{i+1}\ X_{i+2}\ X_{i+3}]^{\mathrm{T}},$$

$$\frac{\mathrm{d}Y}{\mathrm{d}s} = N[3u^2\ 2u\ 1\ 0]M[Y_i\ Y_{i+1}\ Y_{i+2}\ Y_{i+3}]^{\mathrm{T}}. \tag{2.72}$$

The rationale behind GICOV is simple. When the curve is delineating an object (leukocyte in the following example) with somewhat constant edge strength, squared mean intensity μ is large

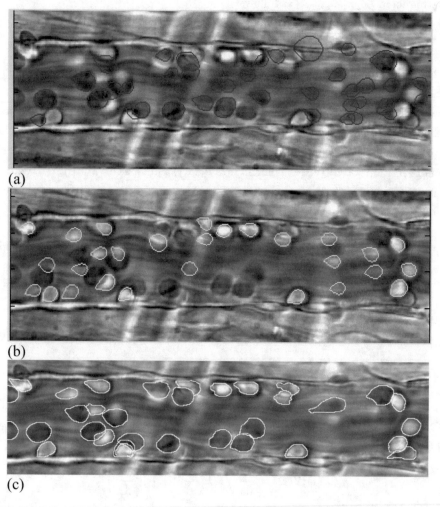

(a)

(b)

(c)

FIGURE 2.18: Segmentation of leukocytes by teardrop B-spline snakes on intravital microscopy images. (a) Teardrop shapes delineating potential leukocytes appearing darker than the background. (b) Teardrop shapes delineating leukocytes with bright appearance. (c) one knot-clamped closed B-spline evolution to segment leukocyte shapes. Taken from Reference [20].

in magnitude, while the variance σ^2 is small. It has been demonstrated that in a highly cluttered environment, GICOV is able to discriminate a leukocyte from noise and clutter [19].

To delineate leukocyte shapes, we first build a teardrop-shaped database. The database has teardrops with different values of A, n, and ω. Given an image, we place these shapes at every pixel locations and compute GICOV score for each. At each pixel location, the maximum GICOV score and the corresponding teardrop shape is found. Next, a threshold on the GICOV score is imposed to sieve out the detected leukocytes. The details of this process can be found in References [19–20]. Figure 2.18a and 2.18b show detection of leukocytes through teardrop shapes by maximizing/minimizing GICOV. Note that leukocytes may appear darker (Figure 2.18a) or brighter (Figure 2.18b) than the background in a microscopy image. Thus, for the former class, the criterion is maximizing the GICOV, while for the latter, the criterion is minimizing GICOV to detect leukocytes.

The teardrop model actually serves as a crude boundary delineator for the leukocyte. We can further refine these boundaries by B-spline snake evolution. We model the teardrop shape as a closed cubic B-spline with both knot ends clamped at the control point (X_1, Y_1). As pointed out earlier, the control point sequence for this B-spline is $(2X_1 - X_2, 2Y_1 - Y_2), (X_1, Y_1), (X_2, Y_2), \ldots, (X_N, Y_N), (X_1, Y_1), (2X_1 - X_N, 2Y_1 - Y_N)$. The spline energy function we consider is GICOV computed over the spline:

$$E_{\text{sp}}(\{(X_i, Y_i)\}_{i=1}^{N}) = -\frac{\mu((\{(X_i, Y_i)\}_{i=1}^{N}))}{s(\{(X_i, Y_i)\}_{i=1}^{N})}, \tag{2.73}$$

where the symbol $(\{(X_i, Y_i)\}_{i=1}^{N})$ merely denotes control points $(X_1, Y_1), \ldots, (X_N, Y_N)$ as arguments of functions. Thus, starting with the teardrop shape yielding maximum (or minimum) GICOV as an initial B-spline, we vary the cubic B-spline parameters $(X_1, Y_1), \ldots, (X_N, Y_N)$ to minimize (or maximize) Equation 2.73. We do this via gradient descent method. Figure 2.18c shows the result of cubic B-spline snake evolution starting from the teardrop shapes.

2.5.4 Rigid Contour Snakes

At first, a rigid contour snake (or rigid active surface) might not seem useful. However, when the segmentation problem involves mating contours or surfaces from multiple modalities, the rigid snakes come into play. They can be used as tools of multimodal registration in this case. Essentially, a segmentation-based registration may be implemented using the rigid snakes or surfaces.

Let's convert our scalar snake evolution equations to a more convenient form found by way of linear algebra. Let's assume that the contour or surface is sampled and represented by a set of N discrete vertices vi. Given an initial estimate of the surface V°, the discrete semi-implicit update procedure for the surface can be written in matrix form [21] as:

$$\mathbf{V}^t = (\mathbf{I} + \tau\mathbf{A})^{-1}\left(\mathbf{V}^{t-1}\right) + \tau\mathbf{F}^{t-1}) \tag{2.74}$$

where $\mathbf{I}$ is the $N \times N$ identity matrix, $\mathbf{V}^t = [\mathbf{v}_0^t, \mathbf{v}_1^t, \ldots, \mathbf{v}_{N-1}^t]^T$ and $\mathbf{F}^t = [\mathbf{f}(\mathbf{v}_0^t), \mathbf{f}(\mathbf{v}_1^t), \ldots, \mathbf{f}(\mathbf{v}_{N-1}^t)]^T$ represent the positions and the external forces of vertices at time t, respectively. Also, τ is the time step, and $\mathbf{A}$ is a $N \times N$ banded sparse matrix representing the internal force. Note that $\mathbf{I} + \tau\mathbf{A}$ is a positive definite banded sparse matrix, and such a linear system is solvable—$\mathbf{V}^{t+1}$ can be updated from $\mathbf{V}^t$ [22].

Now, we have set the necessary background to develop *rigid body surfaces*. Such surfaces are constrained to move according to rigid body transformations. A rigid body transformation involves only a rotation $\mathbf{R}$ and translation $\mathbf{T}$. So, the problem becomes one of evolving a snake/surface that contains only a rotation and translation. In such an evolution, it is important to note that internal energy has no bearing. Then, the update of Equation 2.74 can be simplified using

$$\mathbf{V}^t = \mathbf{V}^{t-1} + \tau\mathbf{F}^{t-1}. \tag{2.75}$$

In such a transformation, the new surface may not satisfy the rigid body transformation constraint.

So, we compute the optimal rigid body transformation between the deformed surface and the predefined surface using least squares estimation [23]. Determining the optimal rigid body transformation at time t, denoted by $(\mathbf{R}^t, \mathbf{T}^t)$, in the least-squares sense is equivalent to minimizing

$$\frac{1}{N}\sum_{i=0}^{N-1}\left\|\mathbf{R}^t \cdot \mathbf{v}_i^{t-1} + \mathbf{T}^t - \mathbf{v}_i^t\right\|^2 \tag{2.76}$$

the solution of which is detailed in Reference [24]. Here, the rotation is given by $\mathbf{R}^t = \mathbf{U}^t\mathbf{S}^t[\mathbf{W}^t]^T$ where $\mathbf{U}^t$ and $\mathbf{W}^t$ are provided by the *singular value decomposition* of covariance matrix $\mathbf{C}^t$, i.e.,

$$\mathbf{C}^t \triangleq \frac{1}{N}\sum_{i=0}^{N-1}\left(\mathbf{v}_i^t - \mu^t\right)\left(\mathbf{v}_i^{t-1} - \mu^{t-1}\right)^T = \mathbf{U}^t \cdot \mathbf{D}^t \cdot [\mathbf{W}^t]^T \tag{2.77}$$

where $\mathbf{U}^t$ and $\mathbf{W}^t$ are orthogonal matrices, and $\mathbf{D}^t$ is a diagonal matrix containing the singular values of $\mathbf{C}^t$ [22]. In the rotation, $\mathbf{S}^t$ is given by

$$\mathbf{S}^t = \begin{bmatrix} 1 & 0 & 0 \\ 0 & 1 & 0 \\ 0 & 0 & |\mathbf{U}^t||\mathbf{W}^t| \end{bmatrix}. \tag{2.78}$$

The translation is given by $\mathbf{T}^t = \mathbf{m}^t - \mathbf{R}^t \cdot \mathbf{m}^{t-1}$ where the mean matrices are calculated using $\mu^t = \frac{1}{N}\sum_{i=0}^{N-1}\mathbf{v}_i^t$ and $\mu^{t-1} = \frac{1}{N}\sum_{i=0}^{N-1}\mathbf{v}_i^{t-1}$.

After obtaining the rigid body transformation $(\mathbf{R}^t, \mathbf{T}^t)$, the new surface is updated:

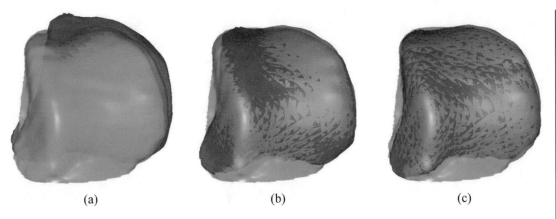

(a)   (b)   (c)

FIGURE 2.19: (a) The initial position of the rigid body surface (red) and its position after (b) 40 and (c) 80 iterations. The target surface (green) is the semiautomatic segmentation result. The alternation of red and green regions in (c) shows that the registration result is successful [25].

$$\mathbf{V}^t = \mathbf{V}^{t-1}\left[\mathbf{R}^t\right]^T + \mathbf{1} \cdot \left[\mathbf{T}^t\right]^T \qquad (2.79)$$

where $\mathbf{1}$ denotes the $N \times 1$ unit vector. Since $\mathbf{V}^t$ is a rigid body transformation of $\mathbf{V}^{t-1}$, $\mathbf{V}^t$ is a rigid body transformation of $\mathbf{V}^0$ given the transitive relationship. Thus, the final result is a rigid body transformation.

An example of a rigid body transformation in which images of the same cartilage surface (taken at different times) is shown in Figure 2.19. The rigid body transformation approach imposes no limitation on the form of the surface. Hence, it is a versatile tool for registration and fusion using segmented surfaces.

CHAPTER 3

Active Contours in a Bayesian Framework

Anytime you have a 50–50 chance of getting something right, there's a 90 percent probability that you'll get it wrong.

—Andy Rooney

3.1 OVERVIEW

Chapter 2 has illustrated active contour segmentation techniques from cost/energy minimization principles. In the present chapter, we are going to broaden the active contour horizon. Here, we cast the active contour technique in a Bayesian framework. The Bayesian framework will help us understand active contours from a probabilistic perspective. Bayesian computation integrates our prior knowledge about a problem into the active contour segmentation. To begin with, we need some background regarding this framework that starts with Reverend Bayes's probability rule defined as:

$$p(A|B) = \frac{p(B|A)p(A)}{p(B)},$$

which essentially follows from the definition of conditional probability on two events A and B. So, we can compute the probability of event A given that event B has occurred if we can compute three quantities: the probability of B given A, the probability of A, and the probability of B. In image analysis, event A is typically representative of an underlying state of an object of interest, while event B represents some characteristic observed in the image or video.

3.2 BAYESIAN FRAMEWORK FOR ACTIVE CONTOUR

3.2.1 An Introduction to the Bayesian Framework

Let us now move to the case at hand: how to use Bayes's rule in active contour computation. Let $(\mathbf{x},\mathbf{y})$ denote an active contour, where $\mathbf{x} = [X_0, X_1, ..., X_{N-1}]^T$ and $\mathbf{y} = [Y_0, Y_1, ..., Y_{N-1}]^T$ are column vectors denoting the snaxel coordinates. In order for Bayes's rule to be applicable here, we need a likelihood or measurement model: $p(G|\mathbf{x},\mathbf{y})$, where G can be a single measurement for the entire active contour or a number of local measurements. Along with the likelihood model, we also need a *prior* probability distribution on the active contour: $p(\mathbf{x},\mathbf{y})$. Next, we can write Bayes's rule to form the *posterior* probability:

$$p(\mathbf{x},\mathbf{y}|G) = \frac{p(G|\mathbf{x},\mathbf{y})\,p(\mathbf{x},\mathbf{y})}{p(G)}. \tag{3.1}$$

We can make use of this posterior probability of the active contour to delineate an object from an image. The desired contour location that should delineate the object is the so-called maximum a posteriori (MAP) estimation from the posterior probability:

$$(\mathbf{x}^*,\mathbf{y}^*) = \arg\max_{\mathbf{x},\mathbf{y}}[p(\mathbf{x},\mathbf{y}|G)]. \tag{3.2}$$

Because the denominator in Equation 3.1 does not involve the active contour, one can ignore it in the MAP estimation. Also, an equivalent criterion for MAP (Equation 3.2) is as follows:

$$(\mathbf{x}^*,\mathbf{y}^*) = \arg\min[-\ln p(\mathbf{x},\mathbf{y}|G)] = \arg\min[-\ln p(G|\mathbf{x},\mathbf{y}) - \ln p(\mathbf{x},\mathbf{y})]. \tag{3.3}$$

We can apply this MAP estimation (Equation 3.3) to the Kass–Witkin–Terzoulopos (KWT) snake model, where the likelihood is provided by:

$$p(G|\mathbf{x},\mathbf{y}) \propto \prod_{i=0}^{N-1} \exp(f(X_i,Y_i)), \tag{3.4}$$

where we assume that G is not a single measurement, but actually N independent measurements made at N snaxel locations. In this case, the measurements are simply the image gradient magnitudes: $f(x,y) = |\nabla I(x,y)|^2 = (\partial I/\partial x)^2 + (\partial I/\partial y)^2$. We further *assume* that the following two statements are true: (1) measurements are conditionally independent given the snaxel locations, and (2) the measurements have an exponential form of probability distribution. Because of the first assumption, we see that the likelihood factors are as in Equation 3.4. The second assumption is made so that we can use Equation 3.3 for MAP estimation. In a sense, both these assumptions are made to simplify our computation. Next, we need a prior probability model for the active contour. For the KWT snake model, the prior probability imposes smoothness in the contour. Thus, our prior belief about the object is that its boundary will be somewhat smooth. This prior can take the following form:

$$p(\mathbf{x},\mathbf{y}) \propto \exp\left(-\frac{1}{2}\mathbf{x}^{\mathrm{T}}A\mathbf{x} - \frac{1}{2}\mathbf{y}^{\mathrm{T}}A\mathbf{y}\right), \tag{3.5}$$

where A is a smoothness matrix given by:

$$A = \begin{bmatrix} c & b & a & & & & a & b \\ b & c & b & a & & & & a \\ a & b & c & b & a & & & \\ & \ddots & \ddots & \ddots & \ddots & \ddots & & \\ & & a & b & c & b & a & \\ a & & & a & b & c & b \\ b & a & & & a & b & c \end{bmatrix},$$

where in turn a, b, and c are as follows: $a = \beta$, $b = -(4\beta + \alpha)$, $c = 6\beta + 2\alpha$. Putting the likelihood (Equation 3.4) and the prior (Equation 3.5) in Equation 3.3, we obtain:

$$(\mathbf{x}^*, \mathbf{y}^*) = \underset{\mathbf{x}, \mathbf{y}}{\arg\min} \left[\frac{1}{2}\mathbf{x}^T A \mathbf{x} + \frac{1}{2}\mathbf{y}^T A \mathbf{y} - \sum_{i=0}^{N-1} f(X_i, Y_i) \right]. \tag{3.6}$$

Note that Equation 3.6 is the same as minimizing the following KWT energy function from Chapter 2:

$$E(X_0, \ldots, X_{n-1}, Y_0, \ldots, Y_{n-1}) = \frac{1}{2}\sum_{i=0}^{n} \alpha(X_{i+1} - X_i)^2 + \alpha(Y_{i+1} - Y_i)^2$$

$$+ \frac{1}{2}\sum_{i=0}^{n} \beta(X_{i+1} - 2X_i + X_{i-1})^2 + \beta(Y_{i+1} - 2Y_i + Y_{i-1})^2 - \sum_{i=0}^{n} f(X_i, Y_i)$$

$$= \frac{1}{2}\mathbf{x}^T A \mathbf{x} + \frac{1}{2}\mathbf{y}^T A \mathbf{y} - \sum_{i=0}^{n} f(X_i, Y_i).$$

Through this exercise, we now know that KWT snake computation can be explained with a Bayesian framework. However, the curious reader is also wondering about what extra mileage we have gained through this process. We have not gained any advantage for the KWT snake in the Bayesian framework; however, we can guess about the flexibility and power of the Bayesian framework at this point. One of the powerful tools in the Bayesian framework is the concept of a prior probability that helps us imbed the prior knowledge in the active contour framework. In the forthcoming sections, we will learn that there is no reason the prior should only be a smoothness prior—we will play with more interesting and meaningful priors. Another form of power and flexibility is the measurement or likelihood that is also at our discretion to be designed.

3.2.2 A Case Study: Mouse Heart Segmentation

In this section, we illustrate a Bayesian snake computation for object delineation. The application is myocardial border extraction from magnetic resonance (MR) images of a mouse heart [26]. Figure 3.1 illustrates such an image showing a mouse heart myocardial border.

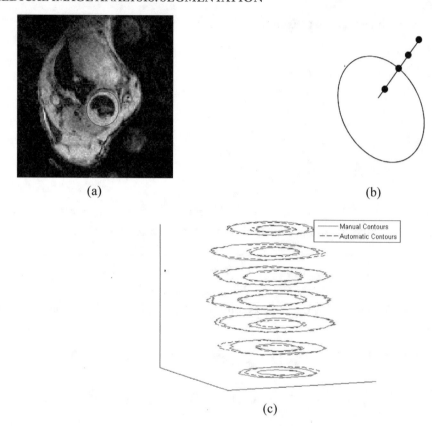

FIGURE 3.1: (a) Endocardial and epicardial borders with initialized active contours. (b) A normal line segment and candidate points on it. (c) Final contours computed by MH sampling-based computation on seven MR slices. Manually drawn contours are also shown for comparison (see [26] for details).

As before, let $\{(X_i, Y_i)\}_{i=0}^{N-1}$ denote N contour points or snaxels. Let $I(x,y)$ be an MR image slice where the myocardium appears. As we have seen before in a probabilistic setting, we think of the snaxels as random variables. The posterior probability of the snaxels is given by:

$$p(\{(X_i, Y_i)\}_{i=1}^{N} | G) = \frac{p(G | \{(X_i, Y_i)\}_{i=1}^{N}) p(\{(X_i, Y_i)\}_{i=1}^{N})}{p(G)}, \tag{3.7}$$

where $p(\{(X_i, Y_i)\}_{i=1}^{N})$ is the prior density and $p(G | \{(X_i, Y_i)\}_{i=1}^{N})$ is the likelihood of the snake $\{(X_i, Y_i)\}_{i=1}^{N}$ based on image measurements G. Unlike the KWT snake here, we have a single measurement G for the entire snake. We call this measurement as gradient inverse coefficient of variation (GICOV) for the snake $\{(X_i, Y_i)\}_{i=1}^{N}$ and is given by:

$$G = \frac{m}{\sqrt{s/N}}, \text{ where}$$

$$m = \frac{1}{N} \sum_{j=1}^{N} [\nabla I(X_j, Y_j) \cdot \vec{\mathbf{n}}(X_j, Y_j)] \tag{3.8}$$

$$s^2 = \frac{1}{N-1} \sum_{j=1}^{N} [\nabla I(X_j, Y_j) \cdot \vec{\mathbf{n}}(X_j, Y_j) - m]^2,$$

In Equation 3.8, $\vec{\mathbf{n}}$ is the outward normal to the active contour. We assume a simplistic image model here—the myocardium is assumed to form step edges of height μ; the image is subsequently corrupted by zero mean Gaussian noise with standard deviation σ_n. For this image model, it has been shown that when the contour is collocated with the object edge, the GICOV has a noncentral Student's t distribution [19]. Thus, the likelihood density $p(G|\{(X_i, Y_i)\}_{i=1}^{N})$ is noncentral Student's t density $\text{NCT}(G; v, \delta)$, where the noncentrality parameter is given as: $\delta = \sqrt{\dfrac{2}{N} \dfrac{\mu}{\sigma_n}}$ and degree of freedom $v = N - 1$.

The prior density is defined as:

$$p(\{(X_i, Y_i)\}_{i=1}^{N}) = p_{\text{initial}}(\{(X_i, Y_i)\}_{i=1}^{N}) p_{\text{smooth}}(\{(X_i, Y_i)\}_{i=1}^{N}),$$

where P_{smooth} is defined:

$$p_{\text{smooth}}(\{(X_i, Y_i)\}_{i=1}^{N}) = \frac{1}{\sqrt{2\pi\sigma_{\text{smooth}}^2}} \exp\left(-\frac{\sum_{i=1}^{N}((X_{i+1} - X_i)^2 + (Y_{i+1} - Y_i)^2)}{2N\sigma_{\text{smooth}}^2}\right),$$

and p_{initial} is defined as:

$$p_{\text{initial}}(\{(X_i, Y_i)\}_{i=1}^{N}) = \frac{1}{\sqrt{2\pi\sigma_{\text{initial}}^2}} \exp\left(-\frac{\sum_{i=1}^{N}((X_i - X_i^0)^2 + (Y_i - Y_i^0)^2)}{2N\sigma_{\text{initial}}^2}\right),$$

where σ's are scaling parameters in the probability density functions. p_{smooth} is a smoothness prior similar to the KWT snake, except for the second-order smoothness term. p_{initial} essentially gives the confidence in the initial snake position $\{(X_i^0, Y_i^0)\}_{i=0}^{N-1}$.

Having defined the likelihood and the prior in Equation 3.7, our task is to determine the MAP, which essentially is the active contour that maximizes the posterior. For the KWT snake, we explored the route of optimization to find the MAP. However, continuous optimization is not always possible or even desirable. For example, if each snaxel is allowed to take only a finite number of possible locations, continuous optimization will be of no use. In such cases, dynamic programming (DP) can be employed (see Reference [27]). Another possibility is the sampling-based technique. Although DP is a good choice for the application here, we will illustrate the sampling-based technique just to inform the reader of a different choice of computational technique for such applications. The sampling-based technique we are going to pursue here is called the Metropolis–Hastings (MH) algorithm [28] (see Acton and Ray, *Biomedical Image Analysis: Tracking* [27] for more information). MH sampling-based MAP estimation has these basic steps: generate samples from the target density (in our case the posterior density) and find out the average or the mode of these generated samples.

To set the stage for the MH algorithm, we first initialize contours on the MR image. In Figure 3.1, we illustrate active contour initializations on the myocardial and the epicardial borders by means of ellipses. A library of ellipses varying in size, orientation, and translation are overlaid on the image. We then determine the GICOV value for each ellipse. The ellipses that maximize and minimize the GICOV value are assigned as the initial contours for the endocardial and epicardial border shown in Figure 3.1a in red and blue, respectively.

After the contour initialization, the initial contour is discretized into snaxels. For each snaxel, we compute the normal to the contour and generate a set of equally spaced K_j candidate points along the normal as shown in Figure 3.1b. For each candidate point, we calculate the outward normal gradient. If the object is darker than the background ($\mu > 0$), we keep only the candidate points corresponding to the local gradient maxima. If the object is brighter than the background ($\mu < 0$), we keep only the candidate points corresponding to the local gradient minima. This reduces the number of candidate points by keeping only the pertinent ones.

The initial ellipse is used as the first sample for the MH algorithm. The algorithm then proceeds by moving along the contour sequentially sampling for each contour point. At a contour point j, one of K_j candidates is selected with probability $p_j(k)$. $p_j(k)$ is calculated by first assigning each candidate point a weight according to:

$$w_i^k = \exp(-(\nabla I(X_{i,k}, Y_{i,k}) \cdot \vec{n}(X_{i,k}, Y_{i,k}) - \mu)^2),$$

where $(X_{i,k}, Y_{i,k})$ is the coordinates of the kth candidate point for the ith snaxel. The probability of sampling a candidate point k for a control point i during the MH algorithm then becomes: $p(X_{i,k}, Y_{i,k}) \equiv p_i(k) = w_i^k / \sum_{m=1}^{K} w_j^m$. The algorithm is described in the algorithm mouse heart MCMC.

Mouse heart MCMC

Initialize

1. Find best ellipse $\{(X_i^0, Y_i^0)\}_{i=1}^N$
2. Pick the first sample as the best ellipse: $S_{x,i}^{(0)} = X_i^0; S_{y,i}^{(0)} = Y_i^0$.
3. For $i = 1, 2, \ldots, N$
4. Compute $p_i(k) = w_i^k / \sum_{m=1}^K w_j^m$, $k = 1, 2, \ldots, K_i$.

Generate samples

1. For $t = 1, 2, 3, \ldots, T$
2. For $i = 1, 2, \ldots, N$
3. Sample $k \sim p_i(k)$
4. Compute Hastings' ratio:

$$r = \frac{p(S_{x,i}^{(t-1)}, S_{x,i}^{(t-1)})}{p(S_{x,i}^{(t)}, S_{x,i}^{(t)})} \frac{p(G|\{(S_{x,i}^{(t)}, S_{x,i}^{(t)})\}_{i=1}^N) p(\{(S_{x,i}^{(t)}, S_{x,i}^{(t)})\}_{i=1}^N)}{p(G|\{(S_{x,i}^{(t-1)}, S_{x,i}^{(t-1)})\}_{i=1}^N) p(\{(S_{x,i}^{(t-1)}, S_{x,i}^{(t-1)})\}_{i=1}^N)}.$$

5. Generate a random variable distributed uniformly in $[0, 1]$: $u \sim U[0, 1]$.
6. If $r < u$, then rollback: $S_{x,i}^{(t)} = S_{x,i}^{(t-1)}; \quad S_{y,i}^{(t)} = S_{y,i}^{(t-1)}$.

   else, assign $S_{x,i}^{(t)} = X_{i,k}; S_{y,i}^{(t)} = Y_{i,k}$.

Compute final contour

$$X_i = \frac{1}{T-B} \sum_{t=B+1}^T S_{x,i}^{(t)}; Y_i = \frac{1}{T-B} \sum_{t=B+1}^T S_{y,i}^{(t)}.$$

The first B samples are highly correlated and can distort the final result significantly. Thus, they are discarded in the algorithm. Typically, $B = 0.2T$. Figure 3.1c shows the results of MH sampling-based cardiac border computation on several MRI slices (taken from Reference [26]).

3.3   ACTIVE MODELS WITH SHAPE PRIORS

In medical and biological object segmentation applications, we often have a priori knowledge about the shape, size, or sometimes the texture of the objects we are segmenting. Naturally, the question arises as to whether it is possible to incorporate this prior information inside active contour or snakes through Bayesian techniques. Active shape models and active appearance models seek to satisfy this prior construction.

3.3.1   Active Shape Models

Suppose the application is segmentation of the myocardium from ultrasound images (see Figure 3.3). The object in this case is deformable. However, the degree of deformation is somewhat limited. This means if we study, say, ten such images and take a "mean" of these ten shapes, then each shape will not be very far from this mean shape. In fact, we can imagine that these shapes have a probability distribution, and we try to learn this distribution from available training cases, for example, the ten shapes. This learned distribution can serve as the prior knowledge that Reverend Bayes first suggested.

How exactly is this prior knowledge about object shape going to help us while performing the object extraction from a test (i.e., never-seen-before) image? If the object extraction computation is a sampling-based statistical inference technique like the one we have learned in the previous section, then we can sample several shapes from this learned prior distribution and compute the likelihood of each of these sampled shapes. Once again, the likelihood could be as simple as the image gradient magnitude averaged over the pixels through which the sampled contour passes. Next, our "inferred" shape might be computed by averaging over these sample shapes weighted by the likelihood numbers. Note that if we did not have any clue about the prior distribution, finding out contour samples might be very difficult, something like shooting in the dark, and often, we will not be able to discover the right object we are seeking.

Active-shape models (ASMs) are essentially based on the aforementioned idea of a distribution of shapes [29]. To elucidate ASMs, we first need to be specific when we talk about shapes. In this context, a two-dimensional shape is nothing other than the active contour itself: a set of ordered sequence of points. For the sake of simplicity, in this section, we will denote $\mathbf{x}$ by the ordered coordinate pairs, i.e., $\mathbf{x} = [X_0, Y_0, X_1, Y_1, \ldots, X_{n-1}, Y_{n-1}]$. Thus, when we refer to a distribution of shapes, essentially, we are referring to the distributions of random vectors $\mathbf{x}$. ASM assumes that the distribution of $\mathbf{x}$ is a multivariate Gaussian:

$$\mathbf{x} \sim \frac{1}{\sqrt{(2\pi)^{2n}|\Sigma|}} \exp\left(-\frac{1}{2}(\mathbf{x} - \mu)^T \Sigma^{-1}(\mathbf{x} - \mu)\right), \qquad (3.9)$$

where μ and Σ are, respectively, the mean vector and the covariance matrix for the shape $\mathbf{x}$, and n is the number shape points (snaxels), $|\ |$ denotes a determinant. Note that the covariance matrix Σ is of size $2n$-by-$2n$.

The mean and the covariance matrix are the parameters of the multivariate Gaussian distribution. In principle, one can learn (i.e., estimate) the mean vector and the covariance matrix from the training instances. Once these parameters are estimated, we can proceed to perform the Bayesian inference for an unknown test image by sampling as described earlier. However, there is a potential problem with this approach. The problem stems in estimating the covariance matrix. Typically, the number of training examples is small compared to the number of points (i.e., snaxels) on the shape. Small sample size adversely affects estimation of the covariance matrix. First, for a small sample size, the covariance matrix becomes singular, making the inference extremely difficult. Moreover, the training samples contain noise. So the points obtained from a training shape are not the true locations; instead, they are true locations plus/minus some small deviations (noise). The effect of this noise can be suppressed if the sample size is large. Collectively, these effects are sometimes referred to as the "curse of dimensionality." In plain words, this means, if the number of parameters you are estimating is large, the required sample size for reliable estimation of these parameters should be large too, and the latter grows at a much faster rate than the former. The curse of dimensionality makes direct parameter estimation infeasible for a large number of parameters. Typically, this is the situation with shapes in many biomedical applications. We will see in a moment that the ASM makes a cunning move to combat the curse of dimensionality.

Let us now take a closer look at the multivariate Gaussian shape distribution (Equation 3.9) with the help of linear algebra. This treatment here will eventually lead to the ASM formulation. Because of the *singular value decomposition*, the shape covariance matrix Σ, which is real, symmetric, and positive definite, can be written as a product of three matrices as: $\Sigma = UDU^{\mathrm{T}}$, where U is a $2n$-by-$2n$ orthogonal matrix, i.e., U is the inverse of U^T and vice versa, and D is a $2n$-by-$2n$ diagonal matrix with positive diagonal entries: $\sigma_1^2 \geq \sigma_2^2 \geq \cdots \geq \sigma_{2n}^2 > 0$, i.e., $D = \mathrm{diag}(\sigma_1^2, \sigma_2^2, \ldots, \sigma_{2n}^2)$. Note that n is the number of points (snaxels) on the shape. The determinant of the covariance matrix is now: $|\Sigma| = |UDU^T| = |U||D||U^T| = \prod_{i=1}^{2n} \sigma_i^2$ Also note that the inverse of the covariance matrix can be written as: $\Sigma^{-1} = UD^{-1}U^T$, where D^{-1} is the inverse of D and is the diagonal matrix: $D^{-1} = \mathrm{diag}(1/\sigma_1^2, 1/\sigma_2^2, \ldots, 1/\sigma_{2n}^2)$. Taking these algebraic facts into account, one can write the multivariate Gaussian shape distribution (Equation 3.9) as:

$$\frac{1}{\sqrt{(2\pi)^{2n}|\Sigma|}}\exp\left(-\frac{1}{2}(\mathbf{x}-\mu)^{\mathrm{T}}\Sigma^{-1}(\mathbf{x}-\mu)\right) = \prod_{i=1}^{2n} \frac{1}{\sqrt{2\pi}\sigma_i}\exp\left(-\frac{(\mathbf{u}_i^{\mathrm{T}}(\mathbf{x}-\mu))^2}{2\sigma_i^2}\right), \quad (3.10)$$

where $\mathbf{u}_i$'s are the columns of the matrix U: $U = [\mathbf{u}_1 \cdots \mathbf{u}_{2n}]$. Thus, the multivariate joint Gaussian distribution is now a product of $2n$ univariate Gaussians. From this expression, it should be clear that the multivariate Gaussian distribution (Equation 3.10) is very sensitive due to those univariate Gaussians with small variances σ_i^2; because, for a sharply peaked Gaussian (i.e., having a small variance), a slight change in the estimated value of σ_i^2 and $\mathbf{u}_i$ can produce a lot of overall change to the distribution. It is no wonder that these sharply peaked Gaussians adversely affect the *entire* joint multivariate Gaussian distribution. This is actually the curse of dimensionality manifested here due to the noise contained in the training sample.

Before answering the question of tackling the curse of dimensionality here as done in an ASM, let us try to understand how an inference would actually work, if we assume a perfect covariance matrix estimate. In that case, to sample a random shape, one would create $2n$ random numbers from $2n$ zero mean univariate Gaussian distributions with variances $\sigma_1^2, \sigma_2^2, \ldots \sigma_{2n}^2$. Let us denote them by $v_1, v_2, \ldots, v_{2n}$, respectively. Then, we would obtain the shape $\mathbf{x}$ by solving the following set of $2n$ linear equations:

$$\mathbf{u}_i^T(\mathbf{x} - \mu) = v_i, \ i = 1, 2, \ldots, 2n.$$

The solution is given by:

$$\mathbf{x} = \mu + U[v_1 \ v_2 \cdots v_{2n}]^T = \mu + \sum_{i=1}^{2n} v_i \mathbf{u}_i. \tag{3.11}$$

Likewise, we could create as many random shapes by sampling, then create some likelihood-weighted average of these random shapes to obtain the delineated object on an unknown test image. We will return to this procedure soon to fully appreciate the ASM. Actually, these linear algebra tricks reveal important insight that will be of some help to understand the ASM later. For now, notice from Equation 3.11 that the vector $\mathbf{x}-\mu$ is represented in a new (orthonormal) coordinate system $[\mathbf{u}_1 \ \mathbf{u}_2 \cdots \mathbf{u}_{2n}]$, which is a rotation of the original canonical coordinates $\mathbf{x}$ in $\Re^{2n}$.

Turning our attention to the central question—what could be a remedial measure to combat the curse of dimensionality—ASMs answer this by a straightforward but elegant strategy: ignore the Gaussian distributions with small variances. This is exactly in alliance with what we have just discussed: the parameters of the sharply peaked Gaussians could not be reliably estimated from a small number of training samples. Typically, ASM retains the Gaussians with m $(\ll 2n)$ largest variances and ignores the remaining $2n - m$ Gaussians with small variances. The shape $\mathbf{x}$ is now distributed as follows:

$$\frac{1}{\sqrt{(2\pi)^m |\Sigma_r|}} \exp\left(-\frac{1}{2}(\mathbf{x} - \mu)^T \Sigma_r^{-1}(\mathbf{x} - \mu)\right) = \prod_{i=1}^{m} \frac{1}{\sqrt{2\pi}\sigma_i} \exp\left(-\frac{(\mathbf{u}_i^T(\mathbf{x} - \mu))^2}{2\sigma_i^2}\right), \tag{3.12}$$

where now the $2n$-by-$2n$ covariance matrix Σ_r is given by:

$$\Sigma_r = [\mathbf{u}_1 \ldots \mathbf{u}_m] \, diag(\sigma_1^2, \sigma_2^2, \ldots, \sigma_m^2)[\mathbf{u}_1 \ldots \mathbf{u}_m]^T.$$

Obviously this is a different multivariate Gaussian distribution compared to Equation 3.10. What is the approximation that the ASM model (Equation 3.12) is actually making? Notice that for a zero-mean Gaussian distribution with a small variance, a random sample would most likely be very close to zero. This approximation model is saying that with probability 1, these values are in fact zeros.

So how can one make the previous inference method work in this case too? Here, we would only need m random numbers v_1, v_2, ..., v_m generated from zero-mean univariate Gaussians with variances $\sigma_1^2, \sigma_2^2, ... \sigma_m^2$. The other $2n - m$ random numbers are all zeros (in fact they are deterministic, not random anymore). Thus, one solves the following sets of linear equations:

$$\mathbf{u}_i^T(\mathbf{x} - \mu) = v_i, \; i = 1, 2, ..., m,$$

$$\mathbf{u}_i^T(\mathbf{x} - \mu) = 0, \; i = m+1, ..., 2n$$

The solution is given by:

$$\mathbf{x} = \mu + U[v_1 \, v_2 \cdots v_m \, 0 \cdots 0]^T = \mu + \sum_{i=1}^m v_i\mathbf{u}_i. \tag{3.13}$$

This is how one sample shape $\mathbf{x}$ can be generated. This process can be repeated as many times as desired.

Although sampling-based inference makes sense in the Bayesian paradigm, one pays a huge computational price here for the present application at hand: the sample size (the number of random shapes) needed to make accurate inference could be large slowing down the computational speed. This increase in computational load is not due to the random shape generation; but to the computation of the likelihood weight. For example, one might need to interpolate image intensity values for each random shape. Interpolation can be an expensive operation. ASMs avoid this classical Bayesian computation for a speedy practical alternative.

The ASM computation is actually saying that a shape $\mathbf{x}$ is a *feasible shape*, if in the rotated coordinate system $[\mathbf{u}_1 \, \mathbf{u}_2 \cdots \mathbf{u}_{2n}]$, the last $2n - m$ coordinates of $\mathbf{x}-\mu$ are all necessarily zeros. Whereas in addition to this requirement, a Bayesian practitioner would say that the first m components are distributed as Gaussians with variances $\sigma_1^2, \sigma_2^2, ... \sigma_m^2$, the ASM is relaxed about these first m components. In fact, the ASM assumes they are uniformly distributed. With these assumptions, we can formulate the ASM as a minimization of a cost function $f(\mathbf{x})$. A familiar example of the cost function is the negative sum of image gradient magnitude on the shape $\mathbf{x}$. This is the same cost function that has been described in conjunction with active contour models in Chapter 2. One would want to minimize this cost function within the feasible solution space, in this case, the space of feasible shapes. Thus, the minimization problem ASM solves is:

$$\min_{\mathbf{x}} f(\mathbf{x}),$$
$$\text{subject to} : \mathbf{u}_i^T(\mathbf{x}-\mu) = 0, \text{ for } i = m+1, m+2, ..., 2n. \tag{3.14}$$

The constraints in this minimization problem merely mean that the solution vectors will be of the form:

$$\mathbf{x} = \mu + U[v_1 \, v_2 \cdots v_m \, 0 \cdots 0]^T = \mu + [\mathbf{u}_1 \, \mathbf{u}_2 ... \mathbf{u}_m][v_1 \, v_2 ... v_m]^T. \tag{3.15}$$

The shape representation (Equation 3.15) is sometimes called *modal* representation. The variables v's are modal parameters and the vectors $\mathbf{u}$'s are the modes of variation. Thus, the original minimization problem is now transformed as:

$$\min_{v_i, i=1,\dots,m} f(\mu + [\mathbf{u}_1,\dots,\mathbf{u}_m][v_1,\dots,v_m]^T). \tag{3.16}$$

The gradient descent on Equation 3.16 with the new variables will be given by:

$$v_i^{new} = v_i^{old} - \tau \left(\mathbf{u}_i^T \frac{df}{d\mathbf{x}} \right), \; i = 1,\dots m, \tag{3.17}$$

where τ is the time-step length, and $\dfrac{df}{d\mathbf{x}}$ is a vector derivative (it is a vector of length $2n$) and is given by:

$$\frac{df}{d\mathbf{x}} = \left[\frac{\partial g(X_0, Y_0)}{\partial x}, \frac{\partial g(X_0, Y_0)}{\partial y}, \frac{\partial g(X_1, Y_1)}{\partial x}, \frac{\partial g(X_1, Y_1)}{\partial y}, \dots, \frac{\partial g(X_{n-1}, Y_{n-1})}{\partial x}, \frac{\partial g(X_{n-1}, Y_{n-1})}{\partial y} \right]^T,$$

where g is the negative image gradient magnitude: $g(x, y) = -|\nabla I(x, y)|^2$.

In terms of the original variable, shape $\mathbf{x}$, the update rules (Equation 3.17) turn out to be a simple update:

$$\mathbf{x}^{new} = \mathbf{x}^{old} - \tau \sum_{i=1}^{m} \left(\mathbf{u}_i^T \frac{df}{d\mathbf{x}} \right) \mathbf{u}_i. \tag{3.18}$$

In principle, one can replace the derivatives of f by gradient vector flow (p, q) as in the snake computation. In that case, the vector $\dfrac{df}{d\mathbf{x}}$ takes the following form:

$$\frac{df}{d\mathbf{x}} = -[p(X_0, Y_0), q(X_0, Y_0), p(X_1, Y_1), q(X_1, Y_1), \dots, p(X_{n-1}, Y_{n-1}), q(X_{n-1}, Y_{n-1})]^T.$$

Now the ASM algorithm takes the following form:

Algorithm ASM. Compute image-based term: $g(x, y) = -|\nabla I(x, y)|^2$.

Initialize the shape: $X_i = \mu_i^x + x^*, \; Y_i = \mu_i^y + y^*, \; i = 0, \dots, n - 1,$

where $(x^*, y^*) = \arg\min_{x,y} \sum_{i=0}^{n-1} g(\mu_i^x + x, \mu_i^y + y)$ and (μ_i^x, μ_i^y) is the mean shape obtained in the training step (see *Training of ASMs*)

Repeat until convergence

1. Compute cost gradient $\dfrac{df}{d\mathbf{x}}$ for the current shape $\mathbf{x}$:

$$\frac{df}{d\mathbf{x}} = \left[\frac{\partial g(X_0, Y_0)}{\partial x}, \frac{\partial g(X_0, Y_0)}{\partial y}, \dots, \frac{\partial g(X_{n-1}, Y_{n-1})}{\partial x}, \frac{\partial g(X_{n-1}, Y_{n-1})}{\partial y} \right]^T,$$

2.   Update shape: $\mathbf{x} = \mathbf{x} - \tau \sum_{i=1}^{m} \left(\mathbf{u}_i^T \dfrac{\mathrm{d}f}{\mathrm{d}\mathbf{x}} \right) \mathbf{u}_i.$

Notice that the initial shape $\mathbf{x}$ is the learned mean shape, $\mu = [\mu_0^x, \mu_0^y, \ldots \mu_{n-1}^x, \mu_{n-1}^y]$, placed on the image with a suitable translation. Note also that this algorithm requires the learning of m orthonormal basis vectors $\mathbf{u}$'s. These vectors and the mean vector are estimated from the training shapes. Some care must be exercised in choosing the time-step parameter τ here. A large value of τ may render the algorithms unstable/oscillatory. A rule of thumb is that choose τ such that: $\tau |\mathbf{u}_i^T \dfrac{\mathrm{d}f}{\mathrm{d}\mathbf{x}}| \leq 3\sigma_i$, for $i = 1, \ldots, m$. The rationale for the rule is that each normally distributed modal parameter is kept within three times its standard deviation.

Let us now turn our attention to Bayes's advice of using the prior knowledge. We have just seen that the sampling-based Bayesian inference is not the best option for ASM. Is there any other faster Bayesian method for the ASM? It turns out that, indeed, there is. In the Bayesian perspective, you will minimize a regularized version of the aforementioned cost function. The regularization takes into account the prior knowledge. The prior knowledge in this case is that the modal parameters v_i's have Gaussian distributions with known mean (zero for all of them) and variances σ_i^2. Recall that the snake energy functional also has a regularization term called internal energy. In the case of ASM, the regularized cost function looks like:

$$\min_{\mathbf{x}} f(\mathbf{x}) + \lambda \sum_{i=1}^{m} \frac{v_i^2}{\sigma_i^2}, \tag{3.19}$$

$$\text{subject to} : \mathbf{u}_i^T (\mathbf{x} - \mu) = 0, \text{ for } i = m+1, m+2, \ldots, 2n.$$

Compare Equation 3.19 with the previous objective function Equation 3.14 and observe that the difference is only the regularization term, which is added to the image-based cost with a user-tuned weighting parameter λ. This objective function can be transformed as before to the following one without constraints:

$$\min_{v_i, i=1, \ldots, m} f(\mu + [\mathbf{u}_1, \ldots, \mathbf{u}_m][v_1, \ldots, v_m]^T) + \lambda \sum_{i=1}^{m} \frac{v_i^2}{\sigma_i^2}. \tag{3.20}$$

We can now minimize this objective function by way of an iterative method. The iterative minimization leads to what we call Algorithm Bayesian ASM:

Algorithm Bayesian ASM
Initialize the shape $\mathbf{x}$.
Compute image-based term g (or GVF).
Repeat until convergence

1.   Compute cost gradient $\dfrac{\mathrm{d}f}{\mathrm{d}\mathbf{x}}$ for the current shape $\mathbf{x}$.

2. Compute: $v_i = \left(\dfrac{-\tau}{1 + \lambda \tau / \sigma_i^2} \right) \mathbf{u}_i^T \dfrac{df}{d\mathbf{x}}$, for $i = 1, \ldots, m$.

3. Update shape: $\mathbf{x} = \mathbf{x} + \displaystyle\sum_{i=1}^{m} v_i \mathbf{u}_i$.

Initial shape can be chosen in the same previous way. Once again, one should ensure that $|v_i| \leq 3\sigma_i$, for $i = 1, \ldots, m$. This algorithm has a faster convergence rate because one can choose a larger time-step size τ without affecting the stability.

Let us point out one very important aspect of these algorithms. The algorithms assume that the initial shape that is placed on the image is sufficiently close to the object to be segmented. Thus, explicit translation, rotation, scaling, or any other such transformation of the shapes are not performed. The external force computed from the image, such as with GVF [6] or VFC [8], drives the shape in a constrained manner. Thus, the deformation is limited here. Cootes et al. describe different versions of these algorithms [29]. In their version, the shape is made to undergo an explicit affine transformation. For some applications, this version may be more suitable, where large rotations/scaling is desired. For a more principled way of integrating affine transformation into shape-constrained snakes, see Reference [30] for a description of affine invariant eigensnakes.

3.3.2   Training of ASMs

To complete the discussion on ASM, we illustrate the training of ASM in this section. In the sequel, we will reveal an extremely important connection: the relationship of ASM with principal component analysis (PCA). First, the meaning of training should be clear in this context. By training of ASM, we mean that we estimate the mean shape μ and the modes of variation $\mathbf{u}_1 \ldots \mathbf{u}_m$.

Given N training shapes $\mathbf{x}_i$'s, one can hardly expect that they are aligned with each other. Thus, the first step in training ASM is the alignment or the registration of the shapes. A variety of ways can be employed for this task. In the section that follows, we present a very intuitive and commonly practiced method.

Aligning of training shapes

1. Choose first shape as the mean shape
2. Register each of the remaining shapes to this one. One can use iterated closed point (ICP [31]) for this registration.
3. If error tolerance is not met, repeat step 2.

After the alignment, the next step is to learn the mean and the modal variations. Let $\mathbf{x}_j = [X_{0,j}, Y_{0,j}, \ldots X_{n-1,j}, Y_{n-1,j}]^{\mathrm{T}}$, $j = 1, 2, \ldots, N$ represent N-aligned training shapes. The mean shape and the covariance matrix are computed as:

$$\boldsymbol{\mu} = \frac{1}{N}\sum_{j=1}^{N} \mathbf{x}_j,$$

$$\Sigma = \frac{1}{N}\sum_{j=1}^{N} (\mathbf{x}_j - \boldsymbol{\mu})(\mathbf{x}_j - \boldsymbol{\mu})^{\mathrm{T}}.$$

Next, the eigenvector–eigenvalue pairs $\{\mathbf{u}_i, \lambda_i\}$, $i = 1, 2, \ldots, 2n$ of the covariance matrix Σ are computed. The eigenvectors are called the modal variations. The estimated Σ is almost invariably singular because the number of training shapes N is usually smaller than the length of the shape vector $2n$, and the rank of Σ never exceeds N. Thus, most of the eigenvalues are zeros. The standard deviations of the Gaussians in Equation 3.12 are computed next: $\sigma_i = \sqrt{\lambda_i}$, $i = 1, \ldots, N$. Having computed $\mathbf{u}_i$'s and the σ_i's, we now need to decide on a value of m, i.e., how many eigenvectors need to be kept for describing ASM. This is done as follows:

$$m = \min\left\{ i : \sum_{k=1}^{i} \sigma_k^2 \geq 0.95 \sum_{k=1}^{N} \sigma_k^2 \right\}.$$

This is to say that the first m eigenvectors capture 95% of the training data variations. Usually, m is much less than $2n$, the length of the shape vectors.

In practice, one should avoid the route of covariance matrix computation and the subsequent eigenvector–eigenvalue computations because the eigenvector–eigenvalue computation takes significant computational time for the $2n$-by-$2n$ matrix Σ. Usually, the number of training shapes N is much less than the length of the shape vectors, $2n$. In such cases, a faster computation is achieved by using the singular value decomposition of the $2n$-by-N-centered data matrix: $K =$

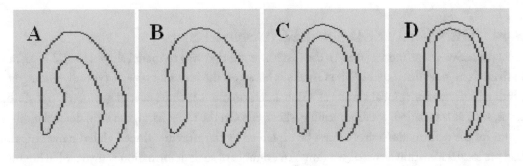

FIGURE 3.2: Panel A: mean shape. B, C, and D: first three modes of variation (eigenvectors) [32].

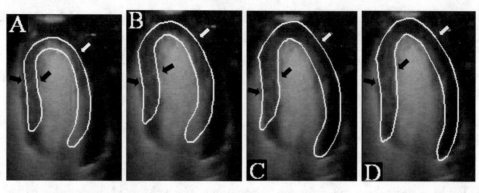

FIGURE 3.3: ASM segmentation. A: initial contour; B, C, D: after 10, 60, and 200 iterations, respectively. Septal wall demarked by dark arrows, lateral epi-cardial border demarked by bright arrows [Pick05].

$K = [(\mathbf{x}_1 - \boldsymbol{\mu}), (\mathbf{x}_2 - \boldsymbol{\mu}), \ldots (\mathbf{x}_N - \boldsymbol{\mu})]$, as $K = ULV^T$ where U is a $2n$-by-N matrix, $L = \text{diag}(l_1, \ldots, l_N)$ is an N-by-N matrix diagonal matrix, and V is an N-by-N matrix. The diagonal entries are non-negative and are usually arranged in the decreasing order: $l_1 \geq l_2 \geq \ldots \geq l_N \geq 0$. It can be shown that the eigenvalues λ_i of the covariance matrix Σ are the square of these diagonal entries: $\lambda_i = l_i^2$, $i = 1, \ldots, N$. Thus, $\sigma_i = l_i$, $i = 1, \ldots, N$. It so happens that the columns of the matrix U are the eigenvectors of Σ. Finally, notice that what we described as ASM training is nothing other than the computation of the principal component analysis of the training shape vectors.

In Figure 3.2, we illustrate the mean and the first three modes of variation of a set of training shapes (see Reference [32] for details). In Figure 3.3, we show the ASM segmentation results. Images from 65 patients were used to create the training database used to segment the image in Figure 3.3. The eigenvector weights were limited to ±1.5 standard deviations of the values observed in training.

3.4   GENERALIZED SNAKES (GSNAKES)

Like the active-shape model (ASM), the Gsnake is another way to model shapes [33]. The name is derived from *generalized snakes*. The Gsnake is based on the idea that objects typically undergo two kinds of deformation: global and local. Global deformations include overall geometric transformations, such as translation, rotation, scaling, shear, etc. On the other hand, the local deformations account for the deformations that cannot be explained by the aforementioned global transformations and are localized within a spatial window. An example of such deformation is myocardial boundary deformation in cine MRI as seen in Figure 3.4. Note that the myocardial borders have undergone global as well as local deformations.

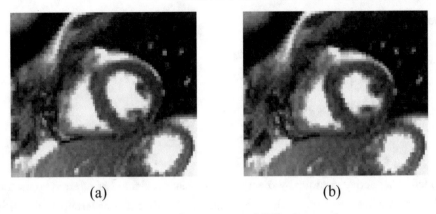

(a)   (b)

FIGURE 3.4: (a) and (b): two consecutive frames in cine MRI of mouse heart.

A characteristic of the Gsnake is that it can model the shape from even a single training shape. Thus, the model could be suitable for the applications where a very limited number of training examples are available. In the Gsnake model, each snaxel is expressed as a linear combination of two of its neighboring snaxels. For example,

$$X_i = \alpha_i X_{i_\alpha} + \beta_i X_{i_\beta}, \text{ and } Y_i = \alpha_i Y_{i_\alpha} + \beta_i Y_{i_\beta}, \qquad (3.21)$$

where the neighborhood indices are chosen as in ([33]):

$$i_\alpha = \begin{cases} i - 1, \text{ for } i > 1 \\ 3, \text{ for } i = 1, \end{cases} \text{ and } i_\beta = \begin{cases} i + 1, \text{ for } i < n \\ n - 2, \text{ for } i = n. \end{cases}$$

Note that the linear combinations (Equation 3.21) are unique, i.e., the coefficients α_i and β_i are unique for an i, if the origin (0,0) and the two points(X_{ia}, Y_{ia}) and (X_{ib}, Y_{ib}) are not collinear. In this formulation, the origin is often taken as one of the snaxels, say the first snaxel ($i = 1$). We can arrange these coefficients in a matrix A (called shape matrix) and form the two shape equations from (Equation 3.21) as follows:

$$A\mathbf{x} = 0 \text{ and } A\mathbf{y} = \mathbf{0}, \qquad (3.22)$$

where $\mathbf{x}$ and $\mathbf{y}$, as before, denote the x and the y coordinates of the snaxels:

$$\mathbf{x} = [X_0\ X_1\ \dots\ X_{n-1}],$$
$$\mathbf{y} = [Y_0\ Y_1\ \dots\ Y_{n-1}].$$

The shape matrix A is defined as:

$$A = \begin{bmatrix} 1 & -\beta_1 & -\alpha_1 & 0 & \cdots & 0 \\ -\alpha_2 & 1 & -\beta_2 & 0 & 0 & \cdots \\ 0 & -\alpha_3 & 1 & -\beta_3 & 0 & \cdots \\ \cdots & \cdots & \cdots & \cdots & \cdots & \cdots \\ 0 & 0 & \cdots & -\alpha_{n-1} & 1 & -\beta_{n-1} \\ 0 & 0 & \cdots & -\beta_n & -\alpha_n & 1 \end{bmatrix} \qquad (3.23)$$

Modeling an active contour by means of the shape matrix A is referred to as the Gsnake model. One remarkable property of the shape matrix A is that it is invariant under affine transformation. For example, say each point (X_i, Y_i) undergoes a global affine transformation K (K is a 2-by-2 affine transformation matrix): $[X_i' Y_i']^T = K[X_i Y_i]^T$.

Even then, the shape equations hold with the same shape matrix A: $A\mathbf{x}' = 0$ and $A\mathbf{y}' = \mathbf{0}$, where $\mathbf{x}' = [X_0' X_1' \cdots X_{n-1}']$, $\mathbf{y}' = [Y_0' Y_1' \cdots Y_{n-1}']$. Moreover, two contours will share the same shape matrix A only if there exists an affine transformation between the two contours. These properties make the object recognition achieved by the Gsnake model theoretically sound.

3.4.1 Training of Gsnakes

Training of the Gsnake model simply involves estimating the matrix A from training example shape(s). Note that one training example is sufficient to compute the matrix A. If, however, more that one training example is available then, like the ASM, one needs to align/register them first. Then, compute the coefficients of the matrix A by the least squares technique. Why least squares? Because, if we have N training contours, then for each pair of unknown coefficients α_i and β_i, we will have $2N$ linear equations. So, we have more equations than unknowns, and the least squares approach is thus appropriate.

3.4.2 Segmentation Using Gsnakes

Once the shape matrix A is learned via training, segmentation with the Gsnake is essentially accomplished by finding object boundaries on an unknown image. Given an image where we should find the object(s) by Gsnake, the first step is to initialize the Gsnake(s) on the image. In the next step, we evolve the initialized Gsnake(s) to lock onto the object boundary. This is much like the original snakes, as well as like most other snake-based object delineation techniques. Automatically initializing the Gsnake could be very difficult in general. If the object that is being sought has intricate shape and boundary details, automatic initialization could indeed be very difficult, unless the underlying image is relatively clutter-free. For a cluttered image, if the object shape is simpler, one can use Hough transform to initialize the Gsnake contour, as in Reference [33]. For now, let us assume that the initialization has been performed, either automatically or semiautomatically.

After the Gsnake initialization, the next task is Gsnake evolution. This is performed in a computation framework similar to that of the original snake. Like the original snake framework, there is an energy function that has two additive components: internal energy and the external/image-based energy. For Gsnakes, a typical energy function is:

$$E_{gsnake}(X,Y,X_0,\dots,X_{n-1},Y_0,\dots,Y_{n-1}) = \sum_{i=0}^{n-1} [w_i\{(X_i - \alpha_i X_{i_\alpha} - \beta_i X_{i_\beta})^2 + (Y_i - \alpha_i Y_{i_\alpha} - \beta_i Y_{i_\beta})^2$$
$$+ (1 - w_i) f(X_i + X, Y_i + Y)], \qquad (3.24)$$

where $w_i \in [0,1]$ is a weighting parameter at the ith snaxel. These weights are tunable parameters of the model. Here, f is the external energy based on an edge detection measure as in the original snake. Usually, f is normalized between [0,1] so that the internal and the external energy terms do not numerically overshadow one another. Note that the internal energy accounts for the deviation of a snaxel position from the linear combination of its neighboring snaxels. The coefficients of the linear combination are known (i.e., learned) from the training shapes. Notice also the presence of a coordinate position variable (X,Y) in the external energy functional. This variable constitutes for a translation term for the entire Gsnake contour. As mentioned before, usually the first snaxel (X_0, Y_0) is set to $(0,0)$. Then, the energy functional Equation 3.24 becomes:

$$E_{gsnake}(X,Y,X_1,\dots,X_{n-1},Y_1,\dots,Y_{n-1}) = w_0\{(\alpha_0 X_{0_\alpha} + \beta_0 X_{0_\beta})^2 + (\alpha_0 Y_{0_\alpha} + \beta_0 Y_{0_\beta})^2\}$$
$$+ (1 - w_0) f(X,Y) + \sum_{i=0}^{n-1} [w_i\{(X_i - \alpha_i X_{i_\alpha} - \beta_i X_{i_\beta})^2$$
$$+ (Y_i - \alpha_i Y_{i_\alpha} - \beta_i Y_{i_\beta})^2\} + (1 - w_i) f(X_i + X, Y_i + Y)].$$

$$(3.25)$$

Minimization of Equation 3.25 is performed in two interleaved iterative steps:

Step 1: Update translation: $X = X + \tau\delta X$, $Y = Y + \tau\delta Y$, where τ is the time-step length, and $(\delta X, \delta Y)$ is the global displacement. Various techniques may be employed to obtain this overall global displacement. One particularly simple way would be to compute average GVF or gradient of external energy force along the Gsnake.

Step 2: Apply dynamic programming (See Reference [27]) to minimize: $E_{gsnake}(X,Y,X_1,\dots, X_{n-1},Y_1,\dots Y_{n-1})$ over the variables $(X_1,\dots, X_{n-1}, Y_1,\dots Y_{n-1})$.

Steps 1 and 2 are repeated until convergence for the Gsnake is reached. One can employ other optimization techniques such as gradient descent for step 2 as well. However, the additive

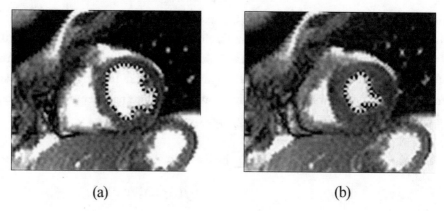

(a)   (b)

FIGURE 3.5: (a) (b) Segmentation of a mouse heart chamber by Gsnake model.

form of the energy functional can be very suitable for dynamic programming, assuming the variables take only some discrete values, such as image grid locations.

The tuning parameters w_i's are typically set all to a single scalar value. However, Lai and Chin [34] also suggests a minimax technique to automatically set these parameters. In the minimax setting, the following energy functional is minimized:

$$E_{\text{gsnake}}(X,Y,X_0,\ldots,X_{n-1},Y_0,\ldots,Y_{n-1}) = \max(\{(\alpha_0 X_{0_\alpha} + \beta_0 X_{0_\beta})^2 + (\alpha_0 Y_{0_\alpha} + \beta_0 Y_{0_\beta})^2\},$$

$$f(X,Y)) + \sum_{i=0}^{n-1} \max(\{(X_i - \alpha_i X_{i_\alpha} - \beta_i X_{i_\beta})^2 + (Y_i - \alpha_i Y_{i_\alpha} - \beta_i Y_{i_\beta})^2\}, f(X_i + X, Y_i + Y)).$$

$$(3.26)$$

Note that the energy functional Equation 3.26 has no tuning parameter at all. It aims to minimize the worse of the two components: internal and external energy terms. Once again, steps 1 and 2 can be employed for the minimization of Equation 3.26. For this energy function, gradient descent cannot be employed in step 2 as such because the energy function is nonsmooth. However, dynamic programming is still a good choice. In Figure 3.5, we show segmentation of cardiac borders in two cine MRI images with the Gsnake model.

CHAPTER 4

Geometric Active Contours

"Make level paths for your feet and take only ways that are firm."

Proverbs 4:26

4.1 OVERVIEW

Geometric active contours have enjoyed a profound impact on biomedical image analysis. The geometric approach brings with it an important advantage: the ability to adjust to the current image topology. In other words, when multiple objects are present, and the number of such objects is not known, the geometric active contour can automatically segment and delineate an arbitrary number of regions.

4.2 LEVEL SETS AND GEOMETRIC ACTIVE CONTOURS

Here, we provide a basic introduction of levels sets and geometric active contours. For a more complete discussion, see References [35–36].

Essentially, the level set approach to image segmentation is based on the fact that the intersection of a smooth three-dimensional surface and a plane yield a closed set of curves. In a level set technique, $\Phi(x, y, t)$ is the height of the surface at position (x, y) and time t. To derive the geometric active contours, we track the positions (x, y) in which

$$\Phi(x, y, t) = 0. \tag{4.1}$$

This is the so-called zero level set, and without loss of generality, this zero level set can represent a set of contours that split, merge, and delineate certain object boundaries.

To update the geometric active contours, we differentiate $\Phi(x, y, t)$ with respect to time. Using the chain rule, we find [36]

$$\frac{d\Phi(x,y,t)}{dt} = \frac{\partial\Phi(x,y,t)}{\partial t} + \nabla\Phi(x,y,t).(x_t,y_t) \qquad (4.2)$$

where (x_t,y_t) is the velocity at point (x,y) on the geometric contour. We take this partial derivative and set it to zero to achieve an Euler–Lagrange equation (representing the steady state of the contour):

$$\frac{\partial\Phi(x,y,t)}{\partial t} + \nabla\Phi(x,y,t) \cdot (x_t, y_t) = 0 . \qquad (4.3)$$

In this level set model, we assume that the surface moves in the normal direction to the surface. If we take velocity (x_t,y_t) and constrain motion in the normal direction, then we can define the speed of the surface as

$$F = (x_t, y_t) \cdot \mathbf{n} \qquad (4.4)$$

for unit normal $\mathbf{n}$. Here, we can take advantage of geometry and note that

$$\mathbf{n} = \frac{\nabla\Phi(x,y,t)}{|\nabla\Phi(x,y,t)|} \qquad (4.5)$$

on the surface $\Phi(x,y,t)$.

Substitution of Equation 4.5 into Equation 4.4 and then into Equation 4.3 gives the classical geometric snake update equation:

$$\Phi_t + F|\nabla\Phi(x,y,t)| = 0, \qquad (4.6)$$

where $\Phi_t = \dfrac{\partial\Phi(x,y,t)}{\partial t}$.

Typically, after an initialization (e.g., a cone-shaped $\Phi(x,y,0)$ with negative height inside an initial region and positive height outside), we can evolve $\Phi(x,y,t)$ by updating each $\Phi(x,y,t)$ for each (x,y) position at time t. Alternatively, we can use a narrow band technique [37] to update just a band around the zero level set contour.

Now, the fundamental engineering design choice is the selection of a suitable F. Often, the speed is chosen to be a function of the curvature and the image gradient magnitude. Given that the curvature can be computed using

$$\kappa = \mathrm{div}\left[\frac{\nabla\Phi(x,y,t)}{|\nabla\Phi(x,y,t)|}\right], \qquad (4.7)$$

we can form a speed by way of

$$F = \frac{\kappa}{\lambda|\nabla I(x,y)| + \varepsilon}. \qquad (4.8)$$

In Equation 4.8, λ is a constant that weights the importance of the image gradient magnitude $|\nabla I(x,y)|$. ε is a small positive constant that prevents divide-by-zero in regions of zero-gradient magnitude.

The speed F in Equation 4.8 depends on curvature, which is akin to the internal energy of the parametric active contour model and upon the image gradient magnitude, which can be equated to the parametric snake external energy. Instead, the speed of the geometric contour can be based on a classification of the image. For example, given a classification of an image into desired objects with $C(x,y) = 1$ and undesired objects with $C(x,y) = -1$, then we can define a speed as

$$F = \kappa - C(x,y). \tag{4.9}$$

The idea of a binary flow will extend this idea using another powerful property of the level set approach—use of region-based statistics.

4.3   BINARY FLOW

Binary flow defines a geometric contour $\mathbf{C}$ that divides the image into two (possibly unconnected) regions, R and its complement, R_c. Yezzi et al. [38] implemented the binary flow by minimizing the cost functional

$$J = -\frac{\lambda}{2}(u - v)^2 + \int_C ds, \tag{4.10}$$

where u is the mean intensity (or some other feature) inside R, and v is the mean inside R_c. A_u and A_v are the corresponding areas of region R and region R_c. λ weights the data term against the curve length term. An update that serves to minimize Equation 4.10 is

$$\mathbf{C}_t = \lambda(u - v)\left[\frac{I - u}{A_u} + \frac{I - v}{A_v}\right]\mathbf{n} - \kappa\mathbf{n}, \tag{4.11}$$

where κ denotes the signed curvature of curve $\mathbf{C}$, and $\mathbf{n}$ is the outward normal to $\mathbf{C}$.

The curve evolution Equation 4.11 is implemented numerically via Osher's level set implementation [39]. In this case, $\mathbf{C}$ is given by the zero level set of the level set function Φ. In this implementation, Φ is positive inside $\mathbf{C}$ and negative outside $\mathbf{C}$.

$$\Phi_t = |\nabla\Phi|\left\{\nabla \cdot \left(\frac{\nabla\Phi}{|\nabla\Phi|}\right) + \lambda(u - v)\left[\frac{I - u}{A_u} + \frac{I - v}{A_v}\right]\right\}. \tag{4.12}$$

In Equation 4.12, the image I is defined only on the zero level set of the function Φ. The solution of the PDE in Equation 4.12 is described in Reference [40].

In the implementation of Equation 4.12, the narrow band method of Reference [37] can be used to achieve the following level set formulation:

$$\Phi_t = \delta_\varepsilon(\Phi)\,|\nabla\Phi|\left\{\nabla\cdot\left(\frac{\nabla\Phi}{|\nabla\Phi|}\right) + \lambda\,(u - v)\left[\frac{I - u}{A_u} + \frac{I - v}{A_v}\right]\right\}, \qquad (4.13)$$

for which $u = (1/A_u)\displaystyle\iint_\Omega I(x,y)H_\varepsilon(\Phi(x,y))\,dx\,dy,$

$$v = (1/A_v)\iint_\Omega I(x,y)[1 - H_\varepsilon(\Phi(x,y))]\,dx\,dy,$$

$$A_u = \iint_\Omega H_\varepsilon(\phi(x,y))\,dx\,dy, \quad \text{and} \quad A_v = \iint_\Omega [1 - H_\varepsilon(\phi(x,y))]\,dx\,dy.$$

In the above, H_ε is the Heaviside function $H_\varepsilon(z) = \dfrac{1}{2}\left[1 + \dfrac{2}{\pi}\arctan\left(\dfrac{z}{\varepsilon}\right)\right]$, and ε is a regularization parameter. Also, $\delta_\varepsilon(z) = d\,H_\varepsilon(z)/d\,z$.

4.4   AREA-WEIGHTED BINARY FLOW

In Reference [41], Yu and Acton showed that the contour location was in error when binary flow is used in conjunction with a smoothing filter. An area-weighted binary flow was proposed to combat this source of segmentation error.

By comparing the first term in Equation 4.10 with the dissimilarity metric in region-merging-based segmentation, Yu and Acton propose that the bias of the mean-difference method can be eliminated by modifying Equation 4.10 as

$$J = -\frac{\lambda}{2}A_u A_v(u - v)^2 + \int_C d\,s. \qquad (4.14)$$

The contour minimizing Equation 4.14 can be computed by

$$\mathbf{C}_t = \frac{\lambda}{A}(u - v)[A_v(I - u) + A_u(I - v) + (1/2)(A_u - A_v)(u - v)]\vec{N} - \mu\kappa\vec{N}, \qquad (4.15)$$

where A is the sum of A_u and A_v. Equation 4.15 can be expressed in terms of a level set formulation as

$$\Phi_t = \delta_\varepsilon(\Phi)|\Delta\Phi|\left\{\Delta\cdot\left(\frac{\Delta\Phi}{|\Delta\Phi|}\right) + \lambda(u - v)[A_v(I - u) + A_u(I - v)\right.$$

$$\left. + \frac{1}{2}(A_u - A_v)(u - v)]. \right. \qquad (4.16)$$

In Reference [41], Yu and Acton have shown that Equation 4.16 is insensitive to smoothing. Unlike the standard binary flow model, the area-weighted binary flow for active contours yields an unbiased contour location no matter how much the image has been smoothed and regardless of the foreground/background size. An example of the area-weighted binary flow segmentation for an ultrasound image of the prostate is shown in Figure 4.1.

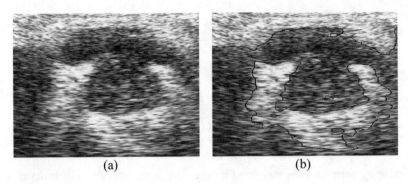

(a)   (b)

FIGURE 4.1: (a) Ultrasound image of prostate (log-compressed data); (b) Contours obtained with the area-weighted model [41].

4.5   ACTIVE CONTOURS WITHOUT EDGES

The "active contour without edges" is an energy functional-based formulation of Chan and Vese [42] that attempts to segment an image into foreground and background regions. The formulation stems from the observation that sometimes there is no clear object boundaries present in an image. Classical edge-based segmentation techniques are almost always bound to produce poor results for these images. Chan and Vese proposed binary image segmentation via minimizing the following region-based energy functional (Chan-Vese or CV functional):

$$F(c_1, c_2, \Phi) = \mu \iint_\Omega \delta_\varepsilon(\Phi(x,y)) |\nabla \Phi(x,y)| \, dx \, dy + \nu \iint_\Omega H_\varepsilon(\Phi(x,y)) \, dx \, dy + \qquad (4.17)$$

$$\lambda_1 \iint_\Omega (I(x,y) - c_1)^2 H_\varepsilon(\Phi(x,y)) \, dx \, dy + \lambda_2 \iint_\Omega (I(x,y) - c_2)^2 (1 - H_\varepsilon(\Phi(x,y))) \, dx \, dy.$$

The first integral in Equation 4.17 penalizes the length of the active contour, the second integral penalizes the area inside the contour, the third and the forth integrals penalize the intensity variances inside and outside the contours. The first two integrals are essentially regularization terms. The symbol H_ε denotes the Heaviside function as before, and δ_ε denotes its derivative. The energy functional Equation 4.17 essentially encodes the classical intraclass-weighted variance minimization into the level set framework.

Minimization of the CV energy functional yields the following three equations that are updated iteratively until convergence:

$$\frac{\partial \Phi}{\partial t} = \delta_\varepsilon(\Phi)[\mu \, \mathrm{div}\left(\frac{\nabla \Phi}{|\nabla \Phi|}\right) - \nu - \lambda_1(I - c_1)^2 + \lambda_2(I - c_2)^2], \qquad (4.18)$$

where μ, ν, λ_1, and λ_2 are non-negative weights.

$$c_1 = \frac{\iint\limits_{\Omega} I(x,y) H_\varepsilon(\Phi(x,y)) \, dx \, dy}{\iint\limits_{\Omega} H_\varepsilon(\Phi(x,y)) \, dx \, dy}, \qquad (4.19)$$

$$c_2 = \frac{\iint\limits_{\Omega} I(x,y)(1 - H_\varepsilon(\Phi(x,y))) \, dx \, dy}{\iint\limits_{\Omega} (1 - H_\varepsilon(\Phi(x,y))) \, dx \, dy}. \qquad (4.20)$$

The similarity between area-weighted binary flow functional and CV functional is not obvious at the first sight. However, a bit of algebraic manipulations reveals that they are in fact equivalent in some sense. Without the two regularization terms, the CV functional is trying to minimize the weighted intraclass variances—intensity variances within the contours and outside the contours. The weights for the variances are simply the respective proportions of the foreground and the background areas. However, it can be shown that the total intensity variance is the sum of weighted intraclass variances and weighted interclass variances. The latter term is, in fact, the negative of the functional in the weighted binary flow. Since the total variance is a constant for an image, minimizing the CV energy functional should actually achieve results very similar to those by weighted binary flow.

4.6 CELL DETECTION USING A VARIATIONAL APPROACH

Many of the same tools that are used in level set methods are applied in variational approaches to image segmentation. Here, we summarize a variational approach to the classical problem of computing a binary segmentation in an inconsistent background.

A basic approach to detecting and segmenting cells would be thresholding the image at some value to separate foreground from background. Such an approach would depend upon the cells having positive contrast with respect to the background. With certain assays that employ fluorescent imaging, for example, this thresholding approach may be feasible. For others, such as with intravital microscopy, the extraction of cells via simple thresholding is impossible.

It may be possible, however, to construct a local thresholding surface that separates the cell segment from the background. The threshold surface u is obtained by interpolation from these values in conjunction with a smoothness constraint. Chan et al. in Reference [43] proposed a variational scheme using level sets for obtaining an adaptive threshold surface. This approach may be considered a variational version of Yanowitz and Bruckstein's algorithm [44].

The approach of [45] is consistent with Chan et al., except that it is not based on edge detection. We observe that pure edge-detection-based algorithms tend to result in less reliable region boundaries. In the spirit of the Mumford–Shah functional [46], we compute B, a continuous edge field that is defined over the entire image domain. The edge function approaches a value of one on the boundaries and approaches zero in non-edge regions. We utilize B to regularize the process of obtaining an adaptive threshold surface u.

Hewer et al., in Reference [47], proposed a suitable boundary function based upon the Mumford–Shah functional [46]. For thresholding surface u and boundary function B, a functional that quantifies success is given by

$$J^{MS}(u, B) = \alpha \int_{\Omega} (u - g)^2 (1 - B)^2 + \beta \int_{\Omega} \Phi(|\nabla u|)(1 - B)^2 + \int_{\Omega} B^2 \qquad (4.21)$$

where g is the image to be segmented. In Equation 4.21, α and β are weighting parameters, and Ω is the two-dimensional domain spanning all supported values of (x, y). The residual between successive estimates of u is given in Reference [47] by

$$r = \alpha(u - g)^2 + \beta(\Phi(|\nabla u|)). \qquad (4.22)$$

With an approximation surface u, the optimal boundary function B can be found explicitly for any nonnegative r:

$$B = \frac{r}{1 + r} \qquad (4.23)$$

and the Mumford–Shah functional reduces to the L_1 norm of the optimal boundary functional B:

$$J^{MS}(u) = \int_{\Omega} \frac{r}{1 + r}, \qquad (4.24)$$

which can be minimized using the gradient descent method.

To compute an energy functional that assesses the quality of the surface u, we assume that such object boundary function B has already been extracted from the image. Given the original image I, the thresholding surface u can be derived by minimizing an objective functional of the form:

$$J(u, B) = \frac{1}{2}\alpha \int_{\Omega} (u - g)^2 B^2 + \int_{\Omega} \Phi(|\nabla u|) \qquad (4.25)$$

where $\Phi(|\nabla u|)$ is a positive increasing function, $|\nabla u|$ is the modulus of the gradient of the approximation surface u, and a is scalar weight. In the functional of Equation 4.25, we have two main terms. The leftmost term enforces consistency with the edge map, and the rightmost term enforces smoothness of u.

By way of the divergence theorem, the Euler–Lagrange equation for J facilitates gradient descent on J is implemented using

$$u_t = \alpha(g - u)B^2 + \nabla \cdot \left(\frac{\Phi'(|\nabla u|)}{|\nabla u|} \nabla u \right). \qquad (4.26)$$

A reasonable smoothness penalty function is given by $f(x) = x^2/2$. Then, the update PDE in Equation 4.26 becomes

$$u_t = \alpha(g - u)B^2 + \Delta u \qquad (4.27)$$

where Δu is the Laplacian of u.

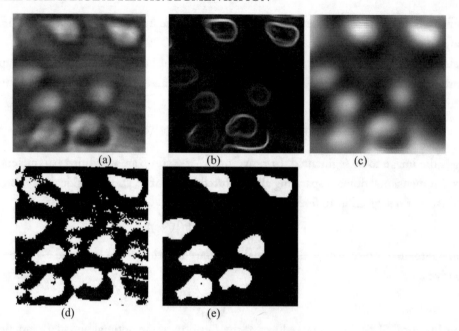

FIGURE 4.2: (a) An intravital video microscopy frame of leukocytes, (b) the boundary function B, (c) the thresholding surface u, (d) the initial thresholding, (e) after isoperimetric and area constraints [45].

Using the surface u, we threshold I to create a binary image. Then, we apply an area constraint and an isoperimetric ratio constraint to identify segments as cells. An example for leukocytes observed in vivo is shown in Figure 4.2.

CHAPTER 5

Segmentation with Graph Algorithms

I like trees because they seem more resigned to the way they have to live than other things do.

—Willa Cather

5.1 OVERVIEW

We have so far discussed the parametric active contour and geometric active contours that are realized via energy minimization with continuous optimization techniques. The focus of this section is to illustrate a few renowned discrete optimization techniques for image segmentation based on *graphs*. Unlike the continuous optimization framework, which typically aims at finding a local minimum in energy, these graph algorithms find globally optimal solutions. Also, the running time for such algorithms is an important consideration.

Before we delve into these fascinating techniques, first let us familiarize ourselves with the proper terminology in this context. If the variables in an optimization method are restricted to take only discrete values, the optimization is called a discrete optimization. When the underlying data structure for a discrete optimization is a graph, we typically refer to it as a graph algorithm. Examples include shortest path, minimum spanning trees, minimum cut computation, and so on. In this section, we will use graph algorithms for two types of computation:

1. Contour or path computation based on shortest path (SP) techniques
2. Region-based segmentation via graph cuts.

5.2 SHORTEST PATH SNAKES

The motivation behind the SP snakes can be explained in terms of user interactive means of object boundary/contour delineation. Suppose we want to quickly delineate an outline in an magnetic resonance spine image. Instead of drawing a closed contour by the mouse cursor, we certainly would like to use a few mouse clicks and expect that the majority of the boundary delineation task will be

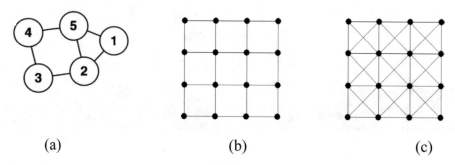

(a)   (b)   (c)

FIGURE 5.1: (a) A graph. (b) Four-connected image grid. (c) Eight-connected image grid.

performed automatically after this user interaction. If a few places of the object boundary contour need correction, we would like to do that at a latter stage, where, again, we do not want to spend much time or effort. This kind of image analysis tasks can be performed quite well with the SP snakes. The same task can also be made automatic with SP snakes by using a trick to be detailed shortly.

Before making some headway into the segmentation algorithm, some requisite preliminaries about graphs and SP are laid out here. A graph G is comprised of two sets—a set of vertices V and a set of edges E. It is commonly denoted as a pair $G = (V, E)$. Figure 5.1a illustrates a graph. For this graph, the vertices are as follows: $V = \{1, 2, 3, 4, 5\}$. The edges in a graph are the connections among the vertices. For the graph in Figure 5.1a, the set of edges is: $E = \{(1, 2), (1, 5), (2, 3), (2, 5), (3, 4), (4, 5)\}$. Figure 5.1b and 5.1c show two graphs on a rectangular image domain. The first one is a four-connected graph and the second one is an eight-connected graph.

A path is a collection of connected edges between a source vertex and a destination vertex. For example, in Figure 5.2, between the source vertex A and destination vertex B, an example path is $\{(A, C), (C, D), (D, B)\}$. Another path is $\{(A, E), (E, D), (D, B)\}$. Further, we can attach weights to the edges in a graph. For the example in Figure 5.2, the numbers labeling edges signify the edge weights. Thus, on an edge-weighted graph, a path will have a weight, too. In the previous example, the path $\{(A, C), (C, D), (D, B)\}$ has a weight $2 + 4 + 1 = 7$, whereas the path $\{(A, E), (E, D), (D, B)\}$ has a weight 5. If these edge weights define some kind of distance between two vertices, then we would interpret that the latter path is shorter than the former. In this sense, we refer to the weight of a path as the length of that path. In a general setting, the length/weight of an edge can be set any real number—positive, negative, or zero. Note that for this example graph in Figure 5.2, we can interchangeably use the terms source and destination, i.e., there is no harm in interchanging the source and the destination vertices as the paths remain the same between them.

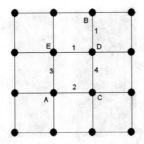

FIGURE 5.2: Paths on a graph and their weight.

In the SP problem, given a source vertex s and a destination vertex t in a graph, we want to find out the shortest path P between them. There are some variations in the shortest path problem. The one stated here is called the point-to-point shortest path. For one of the most recent advances on the point-to-point SP problem, see Reference [48]. The two classical versions of the problem are one source to all destinations, and all source to all destinations. The classical algorithm for one source to all destinations is Dijkstra's algorithm [49], which restricts the weight/length of the edges to be positive.

Let us now turn our attention from the graph and SP preliminaries to the snake computation. We can think of two user chosen points as the source and the destination vertices on the image domain. Now, the problem is to find an SP between the source and the destination points, so that the SP defines the object boundary. It should now be obvious that if the SP needs to be the object boundary, we require an effective edge weight function. For the images shown in Figure 5.3, image gradient magnitude seems to be a dominant cue for delineating the object boundary. Thus, we define the cost here as:

$$D(i, j) = \exp\left(-\frac{|\nabla I(i)| + |\nabla I(j)|}{2w}\right), \qquad (5.1)$$

where i and j are any two adjacent pixels on the image domain. w is a positive user defined parameter. For an eight-connected graph, we need to define these costs for all eight pairs for every interior pixel. For the image border pixels, we define these costs for their available neighbors as well. Figure 5.3 shows the result of interactive segmentation. $D(i, j)$ is essentially a distance between two neighboring pixels i and j. Note that if i and j are object edge pixels, the image gradient magnitude will be large at their locations, and accordingly, the distance between them will be smaller. On the other hand, if both of i and j are not object edge pixels, or if just one of them is not an object edge pixel, then the distance between them will be not be as small. To illustrate, we used Dijkstra's SP algorithm for computing SP in Figure 5.3.

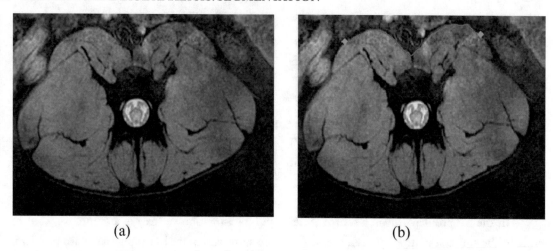

FIGURE 5.3: (a) An MR image slice of spine. (b) Source and destination pixels and the SP between them.

Of course, it is desirable to use the SP approach without user interaction in some scenarios. Here, the source and destination pixels are supplied automatically in some way. Some additional tricks are usually necessary. One approach involves the creation of *barriers*. To elucidate the idea, let us take a quick look at the graph in Figure 5.1a. If we assume all the edges in this graph have the same weight, the SP from vertex 3 to vertex 4 is the direct link (3,4) between them. However, if this edge (3,4) did not exist for this graph, then the SP from 3 to 4 would be {(3,2), (2,5), (5,4)}. Removal of an edge between two vertices thus creates a barrier between them. This same approach can be taken with the graph formed by the image domain grid (Figure 5.1b or 5.1c). For example, as illustrated in Figure 5.4, from the image grid, we have removed the pixels marked by the vertical and horizontal lines. Vertex removal essentially creates barriers in the image grid because all the edges attached to these vertices are also removed. Let us now consider two pairs of pixels (A,B) and (C,D) as shown in Figure 5.4. With the aforementioned barriers, the SP snake computed between A and B is shown in cyan, while that between C and D is shown in yellow in Figure 5.4. These barriers and the points A, B, C, and D are all chosen automatically with some preliminary processing of the MR slice. Note that barrier-induced SP snakes can also be computed in the user interactive setting.

Although the results in Figures 5.3 and 5.4 are impressive, we must discuss the limitations of SP snakes. One severe limitation of the SP paradigm is inherent in the cost function Equation 5.1. The cost/distance is defined only for an edge in the graph, i.e., only for a pair of adjacent pixels in the image domain. If we desired a smoother object contour, we would be prohibited to drift from the existing edges. So, if the image is noisy, without a proper cost function, SP snake can produce

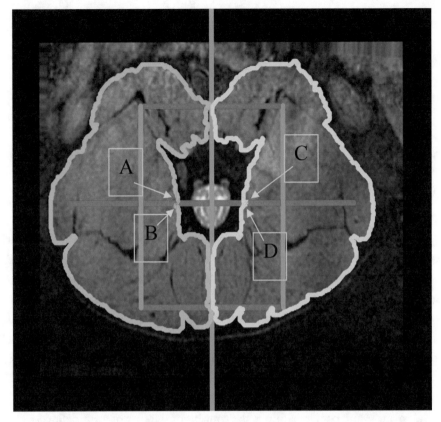

FIGURE 5.4: Barrier-induced SP snake on a MR spine slice. Vertical and horizontal lines represent barriers.

contours with irregularities. However, even in such cases, SP snakes can provide a nice initialization for other snake optimizations such as dynamic programming or gradient descent that can straighten the contours.

5.3 BINARY LABELING WITH GRAPH CUT

Labeling the pixels in the image domain as foreground or background is referred to as binary labeling. We have already seen examples of this in Chapter 4 from a continuous energy minimization perspective. In this section, we will illustrate two well-known discrete optimization approaches, both based on graph algorithms ([50], [51]).

Once again, we will view the image domain as a connected graph, i.e., a collection of connected vertices as shown in Figure 5.1b or 5.1c. The basic principle behind graph-based binary pixel

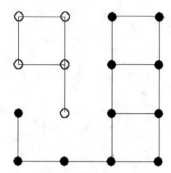

FIGURE 5.5: A cut of a four-connected image grid.

labeling is to divide the image domain into two different sets—label some pixels as 0 (background) and the rest as 1 (foreground). The technical name for this procedure is called *graph cut*. A cut for a graph $G = (V, E)$ with vertex set V and edge set E is formed by removing a subset R of edges from E. This removal of the edges creates two graphs (V_1, E_1) and (V_2, E_2), so that there is no edge between the vertex sets V_1 and V_2. In Figure 5.5, we illustrate a cut of the four-connected image domain grid of Figure 5.1b. Note that Figure 5.5 contains two graphs—one with hollow pixels (say, background) and the other one with solid pixels (say, foreground). These two graphs are the results of a cut, i.e., removal of a set R of edges from Figure 5.1b. The hollow and the solid pixels are merely symbolic and used to denote the binary labeling that resulted from this graph cut.

Central to any graph cut algorithm is the concept of the cost of a cut defined as:

$$\text{cost}(V_1, V_2) = \sum_{(i,j) \in R} W(i, j), \qquad (5.2)$$

i.e., the cost of a cut is the sum of the cost of all the removed edges. There are various ways to obtain a cut of a graph based on the cost of the cut. Here, we will discuss two well-embraced methods—minimum cut (mincut) and normalized cut (Ncut).

5.4   BINARY LABELING WITH MINIMUM CUT

Binary pixel labeling is often performed by minimizing a cost function of the following form:

$$E(p_1, \ldots, p_N) = \sum_{i=1}^{N} f_i(p_i) + \sum_{(i,j) \in \Omega} g_{(i,j)}(p_i, p_j), \qquad (5.3)$$

where $p_1, \ldots, p_N$ are binary pixel labels to be determined, i.e., $p_i \in \{0, 1\}$, N is the total number of pixels in the image domain, f_i is a function that can possibly depend on the pixel location i and model the foreground and the background, $g_{(i,j)}$ is a function that typically imposes smoothness in the solution. For example, $g_{(i,j)}$ will penalize if two neighboring pixels i and j have different labels: $p_i \neq p_j$. Ω is the set of neighboring pixel pairs (i, j). In the set Ω, the order of vertices is not distinguished: the pixel pairs (i, j) and (j,i) are considered the same. Functions of this form (Equation 5.3) are sometimes referred to as a Markov random field (MRF) model [50].

The mincut method can be applied to obtain a globally minimum solution for Equation 5.3 provided the function $g_{(i,j)}$ obeys a *regularity* condition (not to be confused with regularization!) [50]:

$$g_{(i,j)}(0,0) + g_{(i,j)}(1,1) \leq g_{(i,j)}(1,0) + g_{(i,j)}(0,1) \qquad \text{for all neighbors } (i,j) \in \Omega.$$

The trick for minimizing Equation 5.3 via graph cut is to first represent the cost function Equation 5.3 as the cost of a cut of a suitable graph. Then, obtain the minimum cut by efficient graph algorithms (see Reference [50] and references therein). Here, we do not discuss implementations of any minimum cut algorithm; rather, we focus on graph-cut representation of the cost function Equation 5.3.

In what follows is a graph construction procedure adopted from Reference [52]. We build a graph G with $N + 2$ vertices of which N vertices denote the N pixels, and the rest of the two vertices are special ones. One of them is called a source vertex s, and the other one is called a destination/ sink vertex t. Among the N pixel-vertices there are edges between each neighboring pairs (four neighbors or eight neighbors, as the case may be). There are edges from source vertex s to the pixel-vertices. Also, there are edges from the pixel-vertices to the sink vertex t. We refer to such a graph as a source-sink graph. An example source-sink graph is shown in Figure 5.6a, where a small four-pixel (four-connected) image along with a source vertex s and a sink vertex t are seen. We discuss shortly how to set these edge weights in a source-sink graph. A cut on this graph is defined to produce two graphs, so that one graph contains the source vertex s and the pixel-vertices with label $p_i = 0$ and the other graph with a vertex set comprising of the sink vertex t and the pixel-vertices with label $p_i = 1$. In set notations, these two vertex sets are, respectively, $\{s\} \cup \{i : p_i = 0\}$ and $\{t\} \cup \{i : p_i = 1\}$. An example cut of a source-sink graph is shown in Figure 5.6b. The aim here is to set

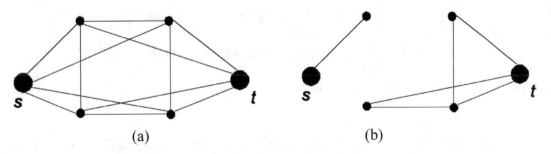

(a)   (b)

FIGURE 5.6: (a) A source-sink graph. (b) A cut of the graph in (a).

the edge weights in the source-sink graph G in such a way that the energy function Equation 5.3 is represented by the cost of a cut of G.

To represent the first component of the energy function Equation 5.3 within the cost of a cut of G, it is written as:

$$\sum_{i=1}^{N} f_i(p_i) = \sum_i [p_i f_i(1) + (1 - p_i) f_i(0)] = \sum_i p_i [f_i(1) - f_i(0)] + \sum_i f_i(0)$$

$$= \sum_i p_i \max(0, f_i(1) - f_i(0)) + \sum_i (1 - p_i) \max(0, f_i(0) - f_i(1))$$

$$+ \sum_i [f_i(0) - \max(0, f_i(0) - f_i(1))]. \tag{5.4}$$

In deriving the last equality, we have used an identity that for any real number z: $z = \max(0, z) - \max(0, -z)$. To represent Equation 5.4 in the source-sink graph G, we add an edge from the source vertex s to the pixel vertex i with weight $f_i(1) - f_i(0)$ if $f_i(1) - f_i(0) > 0$; else, if $f_i(1) - f_i(0) < 0$, we add an edge from i to t with weight $f_i(0) - f_i(1)$. So, if $f_i(1) > f_i(0)$, then there will be an edge (s,i), in G and removal of this edge will cost $f_i(1) - f_i(0)$. On the other hand, if $f_i(1) < f_i(0)$, then there will be an edge (i,t) in G. The cost of removing this edge is $f_i(0) - f_i(1)$. Note that the last summation term in Equation 5.4 does not involve pixel labels and is essentially a constant. Thus, this term need not be represented in the cut cost.

To realize the second term in Equation 5.3, viz., $\sum_{(i,j)\in\Omega} g_{(i,j)}(p_i, p_j)$ within the cut cost, we rewrite it as follows:

$$\sum_{(i,j)\in\Omega} g_{(i,j)}(p_i, p_j) = \sum_{(i,j)\in\Omega} [p_i p_j g_{(i,j)}(1,1) + p_i(1 - p_j)g_{(i,j)}(1,0)$$

$$+ (1 - p_i)p_j g_{(i,j)}(0,1) + (1 - p_i)(1 - p_j)g_{(i,j)}(0,0)]$$

$$= \frac{1}{2} \sum_{(i,j)\in\Omega} (p_i - p_j)^2 [g_{(i,j)}(1,0) + g_{(i,j)}(0,1) - g_{(i,j)}(1,1) - g_{(i,j)}(0,0)]$$

$$+ \frac{1}{2} \sum_{(i,j)\in\Omega} p_i [g_{(i,j)}(1,1) + g_{(i,j)}(1,0) - g_{(i,j)}(0,1) - g_{(i,j)}(0,0)]$$

$$+ \frac{1}{2} \sum_{(i,j)\in\Omega} p_j [g_{(i,j)}(1,1) + g_{(i,j)}(0,1) - g_{(i,j)}(1,0) - g_{(i,j)}(0,0)]$$

$$+ \sum_{(i,j)\in\Omega} g_{(i,j)}(0,0). \tag{5.5}$$

From Equation 5.5, it is clear that we add an edge between every neighboring pixel pair (i,j) with weight $0.5[g_{(i,j)}(1,0) + g_{(i,j)}(0,1) - g_{(i,j)}(1,1) - g_{(i,j)}(0,0)]$. For the second summation term, if $[g_{(i,j)}(1,1) + g_{(i,j)}(1,0) - g_{(i,j)}(0,1) - g_{(i,j)}(0,0)] > 0$, we add an edge from s to i with weight $0.5[g_{(i,j)}(1,1) + g_{(i,j)}(1,0) - g_{(i,j)}(0,1) - g_{(i,j)}(0,0)]$; else, if $[g_{(i,j)}(1,1) + g_{(i,j)}(1,0) - g_{(i,j)}(0,1) - g_{(i,j)}(0,0)] < 0$, we add an edge from i to t with weight $0.5[g_{(i,j)}(0,0) + g_{(i,j)}(0,1) - g_{(i,j)}(1,0) - g_{(i,j)}(1,1)]$. Similarly, for the third summation term, we add an edge from s to j if $[g_{(i,j)}(1,1) + g_{(i,j)}(0,1) - g_{(i,j)}(1,0) - g_{(i,j)}(0,0)] > 0$, with weight $0.5[g_{(i,j)}(1,1) + g_{(i,j)}(0,1) - g_{(i,j)}(1,0) - g_{(i,j)}(0,0)]$; else, if $[g_{(i,j)}(1,1) + g_{(i,j)}(0,1) - g_{(i,j)}(1,0) - g_{(i,j)}(0,0)] < 0$ we add an edge from j to t with weight $0.5[g_{(i,j)}(0,0) + g_{(i,j)}(1,0) - g_{(i,j)}(0,1) - g_{(i,j)}(1,1)]$. If one of these edges exists from previous edge constructions, the existing edge weight is incremented by the current weight. Note that once again nothing needs to be done for the fourth summation term in Equation 5.5 as it is essentially a constant. For a slightly different graph construction, see also Reference [50]. Thus, a cut of G now represents the energy/cost function Equation 5.3 up to an additive constant. Having constructed the source-sink graph G, one obtains the minimum cut, which minimizes the energy function Equation 5.3 and provides the corresponding binary labeling of the image.

Now, we illustrate the mincut on an image shown in Figure 5.7a. Note the cells in Figure 5.7a have essentially two dominant intensity levels, where as the background has practically a single dominant intensity level. Now, we can define the following Gaussian probability models for a pixel i:

$$P(p_i = 0) = \frac{1}{\sqrt{2\pi}\sigma_b} \exp\left(-\frac{(I(i) - \mu_b)^2}{2\sigma_b^2}\right),$$

$$P(p_i = 1) = \frac{0.5}{\sqrt{2\pi}\sigma_{f,1}} \exp\left(-\frac{(I(i) - \mu_{f,1})^2}{2\sigma_{f,1}^2}\right) + \frac{0.5}{\sqrt{2\pi}\sigma_{f,2}} \exp\left(-\frac{(I(i) - \mu_{f,2})^2}{2\sigma_{f,2}^2}\right),$$

where μ_b and σ_b stand respectively for the mean and the standard deviation of the background intensity. The foreground is composed principally of two intensity levels, denoted by $\mu_{f,1}$ and $\mu_{f,2}$, along with respective standard deviations $\sigma_{f,1}$ and $\sigma_{f,2}$. The parameters of these models needs to be

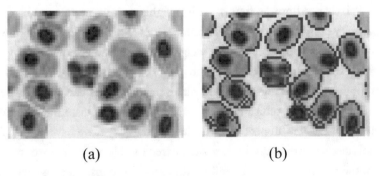

(a)   (b)

FIGURE 5.7: (a) A cell image. (b) Segmentation by mincut.

estimated (say, by maximum likelihood estimation) from training images. Now, we are in a position to define the function f in Equation 5.3:

$$f_i(1) = -\ln(P(p_1 = 1)) \text{ and } f_1(0) = -\ln(P(p_1 = 0)).$$

We can also define the function g in Equation 5.3 as follows:

$$g_{(i,j)}(0,0) = g_{(i,j)} = 0, \text{ and } g_{(i,j)}(1,0) = g_{(i,j)}(0,1) = \lambda$$

where λ is a positive number, essentially a tuning parameter that controls the balance between the two summation terms in Equation 5.3. With such a model, we segmented the image in Figure 5.7a by minimum cut to produce the result shown in Figure 5.7b).

5.5   PIXEL LABELING WITH NORMALIZED CUT

In the previous section, we have seen the action of mincut toward binary labeling. In this section, we will see the act of another graph cut criterion—normalized cut, also known as Ncut. The Ncut criterion has been successfully used in the past for grouping/clustering in computer vision. The basic question grouping/clustering tries to answer is this: given a number of objects N and a number of groups/clusters K, how to put N objects into K groups. The connection to image segmentation should be immediately clear: how to assign labels to image pixels. Obviously, in this setting, one needs to define the affinity or similarity between two objects, in this case, two pixels. The following is an example of affinity between two pixels i and j in an image I:

$$W(i,j) = \exp\left(-\frac{|I(i) - I(j)|^2}{2\sigma_I^2}\right) \exp\left(-\frac{|X(i) - X(j)|^2}{2\sigma_X^2}\right), \qquad (5.6)$$

where $X(i)$ denotes the coordinates of the pixel i in the image domain. So, $|X(i) - X(j)|$ denotes the distance between two pixels i and j. Note that the affinity between two pixels is more when they are closer to each other both in terms of intensity and locations in the image grid. The σ's appearing in Equation 5.6 are scaling parameters that the user must supply. Note that in defining the affinity matrix W, i and j pixels need not be four or eight neighbors. The matrix can be defined for every pixel pair. The affinity matrix W in essence defines a graph for the image domain wherever a pixel is connected with every other pixel. The edge weights in the graph are specified by the elements of W. Such a graph is called a fully connected graph.

Ncut is a grouping/clustering technique where the clustering can typically be performed either in a single shot or in a hierarchical fashion. In the single shot clustering paradigm, one assigns N pixels to K classes or clusters all pixels in one step. When $K = 2$, we have *binary clustering*. In the hierarchical clustering, we essentially form a binary tree of objects by performing repeated binary clustering. Then, the tree pruning can be followed to merge groups. Here, we will only discuss the binary clustering.

One way to perform a binary clustering/labeling is to consider the minimum cut of the fully connected graph induced by the affinity matrix W. However, after playing with an image with mincut on W, one will soon realize that even with a nicely custom-designed cost matrix W, the minimum cut we obtain is quite useless. This is because the minimum cut we typically obtain in such a case is highly imbalanced: we get a very large set of pixels labeled 0 (or 1) and a very small set of pixels labeled 1 (or 0). It might be as bad as one pixel in one set versus the remainder in the other set. This result is not surprising because mincut merely minimizes the total cost of the removed edges. From the segmentation perspective, such imbalanced labeling is of no practical use. A thought that might have left the reader perplexed at this point is why we did not get an imbalanced labeling with minimum cut on the cost function Equation 5.3. The answer to this apparent contradiction is that Equation 5.3 has both an affinity-based term (the second summation) as well as a region-term (the first summation), so the mincut produced acceptable results.

Our naïve effort with minimum cut on the affinity matrix W leads to the following questions: what other meaningful criteria exist to obtain a somewhat balanced cut? and computationally speaking, how efficient are these cuts? The Ncut criterion achieves some balance between the two cuts. Resorting to our previous notations, if a cut of a graph $G = (V, E)$ creates two graphs (V_1, E_1) and (V_2, E_2) by removing a set of edges R from E, then Ncut is defined as:

$$\text{Ncut}(V_1, V_2) = \frac{\text{cost}(V_1, V_2)}{\text{assoc}(V_1, V)} + \frac{\text{cost}(V_1, V_2)}{\text{assoc}(V_2, V)}, \tag{5.7}$$

where 'assoc' is defined as:

$$\text{assoc}(V_1, V) = \sum_{(i,j) \in E_1 \cup R} W(i, j),$$

$$\text{assoc}(V_2, V) = \sum_{(i,j) \in E_2 \cup R} W(i, j).$$

Above, 'assoc' is the association cost of a vertex set V_1 (or V_2) with the original vertex set V. Note that the function "cost" is already defined via Equation 5.2. This Ncut cost Equation 5.7 normalizes a cut cost with respect to the association cost. In fact, it is easy to see that each ratio in Equation 5.7 will never exceed 1; thus, the Ncut value lies between 0 and 2. Normalized cut method minimizes the Ncut cost Equation 5.7 to produce a cut of the fully connected image domain. This cut is essentially a binary labeling of the image.

Let us now examine what happens to the Ncut value when a cut is highly imbalanced. For an imbalanced cut with size of V_1 much larger than that of V_2, $\text{assoc}(V_1, V)$ will be comparable to cost (V_1, V_2), while $\text{assoc}(V_2, V)$ will be much larger than $\text{cost}(V_1, V_2)$, bringing the Ncut value close to 1, which is still an appreciable value, and thus, an unbalanced cut will most likely be not the minimum of Equation 5.7.

With some algebraic manipulations to Equation 5.7, Shi and Malik showed that minimizing Equation 5.7 is equivalent to finding out the eigenvector corresponding to the second smallest eigenvalue of the generalized eigensystem [51]:

$$(D - W)\mathbf{p} = \lambda\,\mathbf{p}, \tag{5.8}$$

where D is a diagonal matrix with diagonal element:

$$D(i, i) = \sum_{j=1}^{N} W(i, j), \ \forall i = 1, \ldots, N.$$

The second smallest eigenvector $\mathbf{p}$ (an N-by-1 column vector) of the system Equation 5.8 provides a minimum of Equation 5.7.

In Figure 5.8a, we show a blood cell image that we want to segment (binary label) with Ncut. The affinity matrix we consider for this problem is:

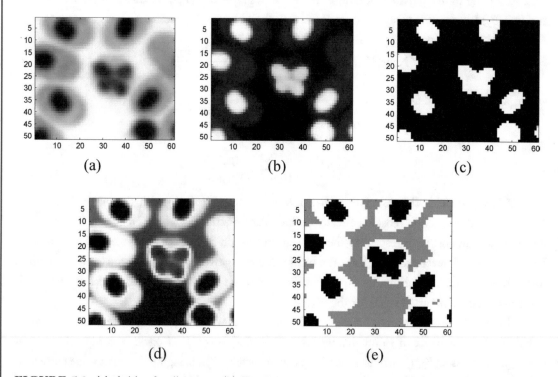

(a)   (b)   (c)

(d)   (e)

FIGURE 5.8: (a) A blood cell image. (b) Eigenvector corresponding to second smallest eigenvalue. (c) Binary labeling via Otsu's method. (d) Eigenvector corresponding to third smallest eigenvalue. (e) Ternary labeling via k-means clustering.

$$W(i,j) = \begin{cases} \exp\left(-\dfrac{|I(i) - I(j)|^2}{2\sigma_I^2}\right) \exp\left(-\dfrac{|X(i) - X(j)|^2}{2\sigma_X^2}\right), & \text{for } |X(i) - X(j)| \le 25. \\ 0, & \text{otherwise.} \end{cases}$$

$$(5.9)$$

Note that W in Equation 5.9 is a sparse matrix, whereas the W in Equation 5.6 is a dense one. Creating a sparse matrix helps speed up the computation of the eigenvector in Equation 5.8. Figure 5.8b shows the second smallest eigenvector **p** of the system Equation 5.8 after we transform the column vector **p** to a two-dimensional matrix by the column-major order. Notice that minimization (Equation 5.7) should lead us to a binary labeling; however, the matrix in Figure 5.8b is not a binary one. So, what happened? It turns out that minimizing Equation 5.7 is *approximately* equivalent to finding out the second smallest eigenvector of Equation 5.8 because there is no guarantee that this eigenvector of Equation 5.8 will be a binary vector. There is a very good computational reason behind this approximation. The reason being that finding a binary solution of Equation 5.7 is an NP-complete problem [51]. For those who are unfamiliar about the notion of NP-completeness [53], it suffices for now to say that finding efficient polynomial time algorithms for these problems has not yet been possible, even with many decades of computer science research. Practically speaking, we can obtain a binary labeling from Figure 5.8b by applying Otsu's thresholding method [54]. This yields the binarized image of Figure 5.8c.

Let us close this discussion by showing that Ncut can, in fact, produce multivalued labeling. For example, we can perform a ternary labeling of the blood cell image in Figure 5.8a by the Ncut method. In this case, along with the second smallest eigenvector, we also need the third smallest one (shown in Figure 5.8d). Next, we need a clustering method such as k-means [55] to produce the ternary labeling shown in Figure 5.8e. The input to the k-means algorithm is both the eigenvectors shown in Figure 5.8b and 5.8d.

· · · ·

CHAPTER 6

Scale-Space Image Filtering for Segmentation

The only basis parameter in computer vision algorithms is scale.

—J.M. Morel and S. Solimini

6.1 OVERVIEW

Often, image filtering is performed to enhance the appearance of an image—to eliminate noise, to deblur, or to enhance contrast. In this chapter, we look at recent filtering methods with a different purpose: to preprocess an image for image segmentation. Such preprocessing methods have shown a significant impact on the problem of segmenting biological and biomedical images.

This chapter on filtering emphasizes filters that produce a *scale space*. A scale space is a family of fine-to-coarse representations of the same signal. Given such a sequence of signals, the proper scale (or scales) for segmentation can be selected. Two types of nonlinear approaches to scale-space generation are highlighted in the chapter. First, we look at diffusion-based processes that utilize a partial differential equation (PDE) framework to adaptively enhance an image. Then, we consider recent advances in morphology.

6.2 SCALE SPACE

To define a scale space, consider a scale parameter d, and a scale space $S(\mathbf{x},d)$ obtained from the original signal $s(\mathbf{x})$, where, in the continuous domain $s(\mathbf{x})$: $R^n \to R$ and $S(\mathbf{x},d)$: $R^n \times R^+ \to R$ [56]. Note that $\mathbf{x}$ is a vector that can represent a position in multidimensional space. The scale parameter d is only allowed to take on non-negative values, where the zero scale indicates the original signal.

Desirable properties of a scale space include fidelity, causality, and Euclidean invariance. We will review each property, in turn, as the properties are important in developing a scale space for the purpose of scale-aware segmentation.

1. Fidelity

    Fidelity enforces the notion that the zero-scale representation is equal to the original, un-filtered signal. Such a requirement can be written as

$$\lim_{d \to 0} [S(\mathbf{x}, d)] = s(\mathbf{x}).$$

(6.1)

where d is a scale parameter particular to the chosen filtering approach. A typical example of a scale-space parameter is found in linear filtering by way of a Gaussian-shaped kernel [57]. In that case, the Gaussian parameter σ (standard deviation) is a suitable scale-space parameter.

2. Causality

    Causality is a desirable property in that it captures the notion that coarse-scale features should be present in a fine-scale representation. That is, each scale-space representation $S(\mathbf{x}, a)$ depends solely on $S(\mathbf{x}, b)$ if $a > b$. A related property is the *monotone* property of *signal features*. The monotone property states that the number of signal features should monotonically decrease as the scale parameter increases. Common features used in image analysis include the zero-crossings of the Laplacian, local signal extrema, or edges.

    In the typical definition of causality, features are allowed to "drift" spatially through scale along continuous paths. The notion of *strong causality* [58] refers to the case in which the spatial position of a coarse feature remains constant until it vanishes at a higher scale.

3. Euclidean invariance

    Obviously, the scale-generating function of the filter should not be sensitive to translation or rotation. A rotated, translated input signal should yield a rotated, translated scale space. Of course, small differences may be produced in actual discrete implementation, but the overall scaling of the image should not be affected by rotation or translation.

    In the remainder of the chapter, we discuss scale-space-generating filters with the aforementioned properties of fidelity, causality, and Euclidean invariance. We argue that scale and the scale-space features are important in biomedical image segmentation. Fidelity to the highest resolution version of the biomedical image is critical, as is the avoidance of spurious features and moving features in scaled version of the image. Finally, our biomedical segmentation methods should not be sensitive to translations or rotations of the specimen or subject.

6.3   ANISOTROPIC DIFFUSION

PDE methods are natural mechanisms for scale-space generation. The solutions to the isotropic *diffusion* or "heat" equation [59]

$$\frac{\partial S(\mathbf{x}, t)}{\partial t} = \Delta S(\mathbf{x}, t),\qquad(6.2)$$

where Δ is the Laplacian, are instances of the Gaussian scale space. The time variable t is related to the scale parameter (standard deviation) of the Gaussian kernel σ by the relation $\sigma^2 = 2t$ [60]. So, instead of successively convolving an image with a Gaussian of greater scale σ, one may implement this scaling via diffusion using Equation 6.2. In this section, we will discuss the advantages of such a diffusion.

The diffusion equation has been generalized to various nonlinear PDEs in order to correct the shortcomings of linear diffusion, such as edge localization. For example, the *anisotropic diffusion* technique of Perona and Malik [59] discourages diffusion in the direction of high-gradient magnitude, where presumably important edges occur. They show that the causality and monotone properties are preserved (with properly chosen diffusion coefficients) when features are defined as local extrema.

The PDE giving a linear scale space (Equation 6.2) can be written as:

$$\frac{\partial S(\mathbf{x}, t)}{\partial t} = \text{div}\left[\nabla S(\mathbf{x}, t)\right],\qquad(6.3)$$

where div is the divergence operator, and ∇ is the gradient. This PDE gives isotropic diffusion, that is, diffusion (smoothing) is equal amounts in all directions and locations. If we devise a function $c(\mathbf{x})$ that takes on high values (near 1) in homogeneous regions and takes on low values near zero at edges, we can inhibit diffusion at edges. Modulating the gradient operator in Equation 6.3 by this diffusion coefficient, we have

$$\frac{\partial S(\mathbf{x}, t)}{\partial t} = \text{div}\left[c(\mathbf{x})\nabla S(\mathbf{x}, t)\right],\qquad(6.4)$$

which gives anisotropic diffusion in the sense that the smoothing can be controlled based on position in the image. Equation 6.4 is the familiar PDE of Perona and Malik [59].

If we take a discrete-domain and discrete-time approximation of Equation 6.4, we obtain

$$[S(\mathbf{x})]_{t+1} - [S(\mathbf{x})]_t = \left[(\Delta T)\sum_{i=1}^{\Gamma} c_i(\mathbf{x})\nabla S_i(\mathbf{x})\right]_t\qquad(6.5)$$

or likewise,

$$[S(\mathbf{x})]_{t+1} = \left[S(\mathbf{x}) + (\Delta T)\sum_{i=1}^{\Gamma} c_i(\mathbf{x})\nabla S_i(\mathbf{x})\right]_t\qquad(6.6)$$

for the "new" iteration $t + 1$ based on the "old" iteration at time t. In Equations 6.5 and 6.6, $c_i(\mathbf{x})$ is the diffusion coefficient in the ith direction, and $\nabla S_i(\mathbf{X})$ is the gradient (directional derivative) in the ith direction. Here, i enumerates the directions of discrete diffusion, e.g., "north," "south," "east," and "west" for a total of Γ directions. Also, in Equations 6.5 and 6.6, ΔT is the time step. Typically, $DT \leq \frac{1}{4}$ for stability.

So, Equation 6.6 can be used to smooth an image and to yield a scale space. Such a scale space is determined by the number of iterations t taken and the particular diffusion coefficient $c(\mathbf{x})$ used. The typical exponential form given by Perona and Malik [59] is

$$c(\mathbf{x}) = \exp\left\{-\left[\frac{\nabla S(\mathbf{x})}{k}\right]^2\right\}, \tag{6.7}$$

where k may be viewed as a soft edge threshold on the gradient. Using such a diffusion that combines Equation 6.6 with Equation 6.7, we can enhance the magnetic resonance (MR) image of an ankle, for example, as shown in Figure 6.1.

Selection of k is ambiguous at best and leads to ad hoc solutions for a given biomedical application. Two other problems with Equation 6.7 are the generation of "staircase" artifacts and the inability to remove impulse noise. To relieve the staircase artifacts (unnatural plateaus in the signal), multigrid approaches [61] have been applied, which are beyond the scope of this discussion.

A regularized diffusion operator can be applied to eliminate impulse noise [62–63]. Here, a filtered version of $S(\mathbf{x})$ is used in the coefficient computation, where

$$c(\mathbf{x}) = \exp\left\{-\left[\frac{\nabla F(\mathbf{x})}{k}\right]^2\right\}. \tag{6.8}$$

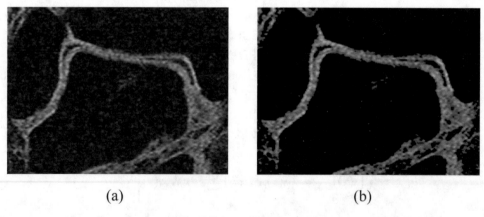

(a) (b)

FIGURE 6.1: (a) A typical coronal MRI slice acquired using the isotropic cartilage sensitive sequence; (b) The same image slice following application of the anisotropic diffusion denoising algorithm. The improvement in definition of the cartilage boundary can clearly be seen [76].

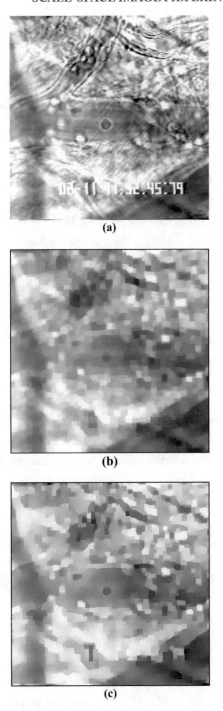

(a)

(b)

(c)

FIGURE 6.2: (a) An image of rolling leukocytes in a mouse cremaster venule; (b) Morphological open–close filter result; (c) Morphological anisotropic diffusion result [77].

In Equation 6.8, $\mathbf{F}$ can be the convolution of $\mathbf{S}$ and a Gaussian kernel with standard deviation σ. Even though Equation 6.8 leads to removal of noise that is preserved in the diffusion via Equation 6.7, this diffusion leads to oversmoothing in addition to the computational load imposed by the additional convolution needed in conjunction with each diffusion iteration.

To eliminate the smoothing of the linear filter, *morphological anisotropic diffusion* can be formed by substituting

$$\mathbf{F} = (\mathbf{S} \circ \mathbf{B}) \bullet \mathbf{B} \tag{6.9}$$

into Equation 6.8, where $\mathbf{B}$ is a structuring element, $\mathbf{S} \circ \mathbf{B}$ is the morphological opening of $\mathbf{S}$ by $\mathbf{B}$, and $\mathbf{S} \bullet \mathbf{B}$ is the morphological closing of $\mathbf{S}$ by $\mathbf{B}$. The efficacy of Equation 6.9 in preserving edges and removing noise within the diffusion process exhibits the power of morphology, which will be expounded on later in this chapter. Consider the example in Figure 6.2. The morphological anisotropic diffusion result of Figure 6.2c gives a clear delineation of the target cells from intravital video microscopy. The standard morphology result in Figure 6.2b suffers a critical loss in image features; in this example, the square (5×5) structuring creates artifacts that are avoided in the point-based diffusion result.

The diffusion coefficient

$$c(\mathbf{x}) = \frac{1}{|\nabla I(\mathbf{x})|} \tag{6.10}$$

is used in mean curvature motion formulations of diffusion [64]. This parameter-free diffusion coefficient is the key to the diffusion highlighted in a subsequent section—*locally monotonic diffusion*. But first, we will highlight a diffusion mechanism that has impacted the world of ultrasound imaging.

6.4   SPECKLE REDUCING ANISOTROPIC DIFFUSION

Application of traditional anisotropic diffusion assumes an additive noise model that exhibits no dependence between noise and signal. This fundamental assumption fails for ultrasound imaging, in which image speckle is dependent upon the signal. In fact, some would argue that speckle is the signal!

For ultrasound images and for radar imagery, filters such as the Lee [65] filter have shown efficacy. This filter provides an adaptive filtering based on the coefficient of variation. The coefficient of variation (local standard deviation over local mean) is a good indicator of discontinuity in speckled images. Remember that the gradient magnitude was our fundamental measure of edge presence in the traditional anisotropic diffusion paradigm. So, our goal in developing a diffusion mechanism for ultrasound is based on using the coefficient of variation in the diffusion coefficient.

In Reference [66], we have shown that the Lee filter is a discrete isotropic diffusion based on the coefficient of variation. So, as anisotropic diffusion adds a diffusion coefficient to the averaging of isotropic diffusion (in the additive noise case), we can construct a speckle reducing anisotropic diffusion (SRAD) by adding a diffusion coefficient to the isotropic smoothing of the Lee filter.

Given the work done by Lee and others, we know that (1) the coefficient of variation is an effective edge detector for speckled imagery, and (2) there exists a baseline value for this statistic. The baseline value is the coefficient of variation in a homogeneous region, also known as a region of fully developed speckle. This coefficient of variation value, called q_0 here, remains constant in homogeneous regions. Therefore, we can construct a diffusion coefficient that essentially compares the coefficient of variation q to q_0, avoiding an ad hoc threshold parameter in the diffusion coefficient. This diffusion coefficient is given by

$$c(q) = \frac{1}{1 + [q^2 - q_0^2]/[q_0^2(1 + q_0^2)]}. \tag{6.11}$$

Unfortunately, the coefficient of variation q is computed in a region, not at a point. But, a PDE as given in Equation 6.4 is a point operation. So, we have derived what we call the *instantaneous coefficient of variation* (ICOV) from the coefficient of variation statistic used in the Lee filter. The ICOV is a discrete-domain operator suitable for inclusion in an anisotropic diffusion process for digital images. For an image intensity S, the ICOV is given by

$$q = \sqrt{\frac{\left(\frac{1}{2}\right)(|\nabla S|/S)^2 - \left(\frac{1}{16}\right)(\nabla^2 S/S)^2}{\left[1 + \left(\frac{1}{4}\right)(\nabla^2 S/S)\right]^2}}. \tag{6.12}$$

The combination of Equations 6.6, 6.11, and 6.12 lead to the full description of SRAD. Equation 6.12 gives a useful edge detector for ultrasound images; this property is explored in Reference [67].

Note that medical ultrasound images are typically viewed using logarithmic compression (to compress the dynamic range of the signal). Unfortunately, these images are often stored in this log-compressed range. Therefore, care must be taken in applying the SRAD algorithm that assumes processing in the intensity domain.

The SRAD approach can be extended to three-dimensional ultrasound [68]. See Figure 6.3 for an example of SRAD for enhancing a mouse heart prior to segmentation of the left ventricular cavity.

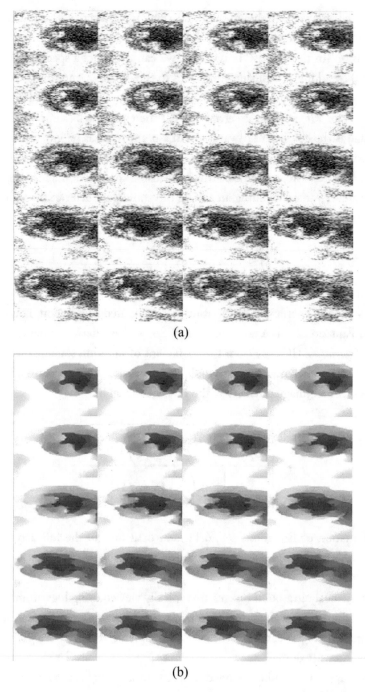

FIGURE 6.3: (a) A sequence of two-dimensional ultrasound images of a murine heart; (b) Three-dimensional SRAD results.

6.5   LOCALLY MONOTONIC DIFFUSION

Smoothness of an image has always been an ill-defined, difficult-to-quantify characteristic. The typical mathematical definitions of continuity and minimized gradient or minimized total variation often penalize signals with strong discontinuities that are otherwise smooth. In the 1980s and 1990s, Restrepo and Bovik [69] pioneered the concept of *local monotonicity*. For one-dimensional signals, the definition of local monotonicity was straightforward: A digital one-dimensional signal is locally monotonic of degree d (LOMO-d) if every contiguous subsequence of length d is either nonincreasing or nondecreasing. Thus, LOMO-d signals consist of ramp edges, step edges, and flat regions. A LOMO-d signal precludes impulses from noise and small-scale features by prohibiting a signal rise and fall within each d-sample window.

Restrepo and Bovik formed smooth signals that allowed both gradual transitions in intensity and abrupt changes in intensity using the tool of *regression* [69]. Essentially, such a regression yields the closest member of the LOMO-d set of signals to a given input signal $s(x)$. A series of regressions with varying values of the scale parameter d, in fact, produce a scale space. Such a scale space could then be used to preprocess the signal for segmentation at a specified scale. The drawback of the regression approach is computational complexity and lack of a clear-cut generalization to multidimensional signals (i.e., images) [70].

Given the attractive definition of local monotonicity in terms of its ability to quantify smoothness and scale without penalizing step edges, we attempted to produce both a one- and two-dimensional filtering scheme based on anisotropic diffusion that computed locally monotonic signals [71].

One negative aspect of diffusion is the arbitrariness of selecting the scale (by stopping at time t). The goal of locally monotonic (LOMO) diffusion is to converge to a LOMO-d signal for desired scale d. Typical diffusion schemes converge to a meaningless constant or zero-curvature signal that is devoid of features. Instead, we seek a method that can specify exactly the feature scale.

Let's consider a one-dimensional signal $S(x)$. On a one-dimensional signal, the basic LOMO diffusion operation is defined using the coefficient Equation 6.10. We then substitute Equation 6.10 into Equation 6.6, noting that $\dfrac{\nabla S_i(\mathbf{x})}{|\nabla S_i(\mathbf{x})|} = \mathrm{sgn}\,[\nabla S_i(\mathbf{x})]$, and using the diffusion coefficient Equation 6.10, yielding

$$[S(x)]_{t+1} \leftarrow \left(S(x) + \frac{1}{2}\,\{\mathrm{sgn}\,[\nabla S_1(x)] + \mathrm{sgn}\,[\nabla S_2(x)]\} \right)_t, \qquad (6.13)$$

with two directions ($\Gamma = 2$) for left and right, and with a time step of $\Delta T = 1/2$ is used. Equation 6.13 is modified for the case where the simple difference $\nabla S_1(x)$ or $\nabla S_2(x)$ is zero. To avoid a denominator $|\nabla S_1(\mathbf{x})|$ of zero, we assign $\nabla S_1(x) \leftarrow -\nabla S_2(x)$ in the case of $\nabla S_1(x) = 0$, and $\nabla S_2(x) \leftarrow -\nabla S_1(x)$ when $\nabla S_2(x) = 0$.

LOMO diffusion produces a LOMO signal via a series of nested diffusions in which the sample spacing for the differences is varied. So, in the computation of the directional derivatives (differences in the discrete-domain case), we use variable spacing between the samples used. Thus, $\nabla S_1(x) = S(x - h_1) - S(x)$, and $\nabla S_2(x) = S(x + h_2) - S(x)$. Then, we define the fixed point of Equation 6.13 in terms of $S(x)$ as $\mathrm{ld}(\mathbf{I}, h_1, h_2)$. If $\mathrm{ld}_d(\mathbf{S})$ denotes the LOMO diffusion sequence that yields a LOMO-d (or greater) signal from input $\mathbf{S}$, then, for odd values of $d = 2m + 1$,

$$\mathrm{ld}_d(\mathbf{S}) = \mathrm{ld}(\ldots\mathrm{ld}(\mathrm{ld}(\mathrm{ld}(\mathbf{S},m,m),m - 1,m),m - 1,m - 1)\ldots,1,1). \qquad (6.14)$$

With the nested sequence in Equation 6.14, the first diffusion implemented is $\mathrm{ld}(\mathbf{I}, m, m)$. Then, diffusion continues with spacings of decreasing widths until $\mathrm{ld}(\mathbf{I}, 1, 1)$ is implemented. For even degrees of local monotonicity $d = 2m$, the sequence of operations is similar:

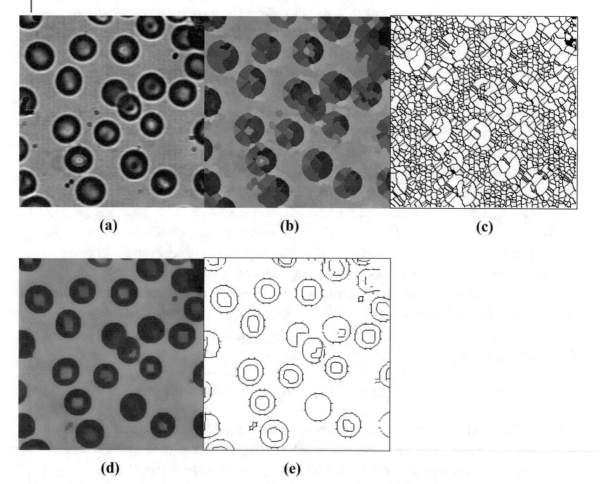

(a)   (b)   (c)

(d)   (e)

FIGURE 6.4: (a) In vitro image of cells; (b) Watershed segmentation; (c) Edges from watershed segmentation; (d) LOMO-3 result from two-dimensional LOMO diffusion; (e) Boundaries from concave-up LOMO segmentation [72].

$$1d_d(\mathbf{I}) = 1d(\dots 1d(1d(1d(\mathbf{I},m,m),m-1,m),m-2,m-1)\dots,1,1). \qquad (6.15)$$

Now, these diffusion operations yield LOMO-d signals for one-dimensional inputs. To extend this method to two dimensions, the same procedure may be followed using four directions ($\Gamma = 4$) instead of two. A second two-dimensional implementation involves diffusing orthogonal to the gradient direction at each point in the image, using the one-dimensional LOMO diffusion. An example of this type of LOMO diffusion is shown in Figure 6.4d. More details on these extensions are given in Reference [71].

6.6   LOCALLY MONOTONIC SEGMENTATION

Achieving a locally monotonic signal by LOMO diffusion leads directly to a segmentation. Acton and Bovik [72] have introduced three methods of defining a segmentation given a LOMO image from a LOMO scale space. They are the *concave-down*, *concave-up*, and *ramp-center* segmentation definitions. With a concave-down segmentat ion, bright objects are delineated. A concave-up delineates dark objects. The ramp-center segmentation is the unbiased alternative, marking the centers of ramp edges as boundaries.

A segmentation of an image can be defined by region boundaries—the edges. For a one-dimensional signal $S(x)$, we define the edges for each of three segmentation approaches.

Concave-down segmentation. $S(x)$ is an edge of a concave-down segmentation if:

1. $S(x)$ is the last (the rightmost) sample of a constant segment that contains C samples: $S(x) = S(x-1), \dots, S(x) = S(x-C+1)$ but $S(x) \neq S(x+1)$.
2. The constant segment is preceded by a decreasing segment and is followed by an increasing segment: $S(x) < S(x-C)$ and $S(x) < S(x+1)$.

or

1. $S(x)$ is the first (the leftmost) sample of a constant segment that contains C samples: $S(x) = S(x+1), \dots, S(x) = S(x+C-1)$ but $S(x) \neq S(x-1)$.
2. The constant segment is preceded by a decreasing segment and is followed by an increasing segment: $S(x) < S(x-1)$ and $S(x) < S(x+C)$.

Concave-up segmentation. $S(x)$ is an edge of a concave-up segmentation if:

1. $S(x)$ is the last (rightmost) sample of a constant segment that contains C samples: $S(x) = S(x-1), \dots, S(x) = S(x-C+1)$ but $S(x) \neq S(x+1)$.

2. The constant segment is preceded by an increasing segment and is followed by a decreasing segment: $S(x) > S(x - C)$ and $S(x) > S(x + 1)$.

or

1. $S(x)$ is the first (leftmost) sample of a constant segment that contains C samples: $S(x) = S(x + 1), \dots , S(x) = S(x + C - 1)$ but $S(x) \neq S(x - 1)$.
2. The constant segment is preceded by an increasing segment and is followed by a decreasing segment: $S(x) > S(x - 1)$ and $S(x) > S(x + C)$.

For two-dimensional signals, we define the edges using the same one-dimensional definitions along rows and columns of the image. For an example of a ramp-center segmentation, see Figure 6.4.

The concave-down and concave-up segmentations are biased with respect to light-on-dark contrast or dark-on-light contrast, respectively. The following ramp-center definition has no such bias.

Ramp-center segmentation. $S(x)$ is an edge of a ramp center segmentation if:

1. $S(x)$ is the center sample of a ramp edge that contains C samples, where C is odd: $S(x - (C - 1)/2 - 1) = S(x - (C - 1)/2 - 2)$, $S(x + (C-1)/2 + 1) = S(x + (C - 1)/2 + 2)$, and the samples are not equal between $(x - (C - 1)/2)$ and $(x + (C - 1)/2)$.

or

1. $S(x)$ is the (left) center sample of a ramp edge that contains C samples, where $C \geq 0$ is even: $S(x - C/2) = S(x - C/2 - 1)$, $S(x + C/2 + 1) = S(x + C/2 + 2)$, and the samples are not equal between $(x - C/2 + 1)$ and $(x + C/2)$.

Here, two-dimensional segmentation is generated by applying the above definition along each row and column.

6.7   MORPHOLOGICAL LOCAL MONOTONICITY

Image morphology is based on two simple operations, erosion and dilation, which are, respectively, the local infimum and supremum operators within a window. Bosworth and Acton showed that local monotonic signals could be defined and generated using a concatenation of these efficient operators [56].

Important morphological filters are formed as combinations of dilation $\oplus$ and erosion $\ominus$. The open operator is defined by

$$A \circ B = (A \ominus B) \oplus B \qquad (6.16)$$

and its dual, the close operator by

$$A \bullet B = (A \oplus B) \ominus B. \qquad (6.17)$$

In Equations 6.16 and 6.17, B is a structuring element—the window that defines the extent of the neighborhood in which erosion and dilation are computed.

It has been known for some time that open and close bound the result of the one-dimensional median filter (from below and above, respectively), and it is also known that the root of the one-dimensional median filter is indeed locally monotonic. With these two facts in mind, we define a locally monotonic signal by the root of

$$S(\mathbf{x}) \leftarrow \frac{S(\mathbf{x}) \circ B + S(\mathbf{x}) \bullet B}{2}. \qquad (6.18)$$

As one may observe, the result is simply the arithmetic average of the open and close operations computed iteratively. The root (the fixed point achieved by successive iteration) of Equation 6.18 defines a LOMO-B signal, in which such a signal may be one-dimensional, two-dimensional, etc. This multidimensional generalization represents a significant difference with the median filter, which can have convergence problems reaching a root in two dimension.

The locally monotonic (LOMO) filter of Equation 6.18 can be used to generate a scale space in which the size of B is increased by increasing the radius of a circular B. The scale space can then be applied to a segmentation problem.

6.8 INCLUSION FILTERS

In this book, we explore level sets to generate geometric contours for multi-object segmentation. Here, in this section, we discuss level sets in the light of describing a grayscale image by a series of thresholded binary images. In fact, the grayscale image of K intensity levels can be represented by $K-1$ level sets via threshold decomposition. Each level set contains a set of connected binary blobs that we call the *connected components*. The boundaries of these connected components are called *level lines*. We show in Chapter 4 that an analysis of these level lines can be used for object detection in biomedical applications.

Assume an integer-valued image $I(x, y)$ have K integer intensity levels, $I(x, y) \in \{0, 1, \ldots, K-1\}$. Then for $\lambda \in \{0, 1, \ldots, K-1\}$, we have the definition for a level set I_λ:

$$I_\lambda(x,y) = 1_{I(x,y) \geq \lambda} \tag{6.19}$$

where $1_{(I(x,y))}$ is the indicator function (and equals one if the subscripted condition is true; equals zero otherwise). The reconstruction of the original image I is achieved via stacking of image level sets I_λ:

$$I(x,y) = \max\{\lambda : \lambda \in \{0,1,\ldots,K-1\}\ I_\lambda(x,y) = 1\}. \tag{6.20}$$

We can define effective filters using image level sets [73–74]. A *connected filter* is defined by a filter that only removes or preserves connected components in their entirety. One such connected filter is the *area-open* filter, written as $I \,\hat{o}\, a$ for image I and area a. An area-open operator removes connected components of area less than the prespecified minimum area from all the level sets of the image. In a complementary fashion, the *area-close* operator ($I \,\hat{\bullet}\, a$) performs identical operation on the complements of the level sets and thus removes dark objects of insufficient area. The sequential concatenation of area open followed by area close for area a is known as *area-open-close* operation: $I \,\hat{o}\, a \,\hat{\bullet}\, a$.

The inclusion filter is a kind of connected operator that can play a useful role in biomedical image segmentation. To define an inclusion filter, we need a few elements from digital topology. One of them is an adjacency tree/forest, which we illustrate here with Figure 6.5 [75]. Figure 6.5a shows a binary image with foreground (black) and background (white) connected components. An

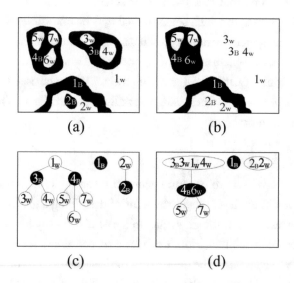

(a)   (b)

(c)   (d)

FIGURE 6.5: (a) A binary image showing foreground- and background-connected components. (b) Result of inclusion filter on the image (a). (c) Adjacency forest corresponding to image (a). (d) Adjacency forest corresponding to the (b) [75].

adjacency forest corresponding to this binary image is shown in Figure 6.5c. The adjacency forest essentially captures the containment relationships among connected components. For example, the connected component 4_B surrounds connected components 5_W, 7_W, and 6_W. Thus, the latter components appear as children nodes of 4_B. Note that the connected components 1_B and 2_W are touching the boundary of the image; they are not surrounded by 1_W.

Next, the concept of a *hole* is central in defining an inclusion filter. Let us first informally illustrate holes by an adjacency forest. For example, consider the connected sets 1_W and 2_W in Figure 6.5c. There are two holes of 1_W, viz., $3_B \cup 3_W \cup 4_W$ and $4_B \cup 5_W \cup 6_W \cup 7_W$. The connected set 2_W has only one hole: 2_B. To formally define a hole, let us denote by $F(S)$ the set of pixels belonging to a subtree rooted at the node S of an adjacency forest. If any node S in the adjacency forest has children nodes $S_1, S_2, ..., S_n$, then we say S has n holes, denoted by $F(S_1), F(S_2),..., F(S_n)$. Using this notion, we define the *filling of a hole $F(S_i)$ of S* as assigning the color of S to $F(S_i)$. In terms of the adjacency forest, filling of a hole refers to merging a child subtree to its parent.

An inclusion filter attempts to fill every hole in a binary image, provided the hole meets a set theoretic criterion, which is a Boolean function on a hole yielding 0 (denoting "fill the hole") or 1 (denoting "retain the hole"). As an example, if the criterion is 1 for a hole if it intersects either 5_W or 7_W and 0 otherwise, then the result of inclusion filter is illustrated in Figure 6.5b. The corresponding adjacency forest is shown in Figure 6.5d. Note that the inclusion filter fills only those holes for which the criterion yields 0 and retain all the other holes. There is an exception to this rule: if a hole touches any image border, then the inclusion filter retains it, irrespective of the value yielded by the criterion associated with this component.

Not all criteria are permissible in defining inclusion filters. Only increasing criteria can define inclusion filters [75]. We call a Boolean criterion T increasing if for two sets of pixels S_1 and S_2, $T(S_1) \geq T(S_2)$, whenever $S_1 \supset S_2$. For the application at hand, we have a seed point $\mathbf{x}$ to begin our filtering. The Boolean criterion for a set S is as follows:

$$T_{\mathbf{x}}(S) = \begin{cases} 1, & \text{if } \mathbf{x} \in S, \\ 0, & \text{otherwise.} \end{cases} \tag{6.21}$$

One can easily show that criterion Equation 6.21 is increasing. An inclusion filter that uses Equation 6.21 is called a marker inclusion filter. If any hole S touches the image boundary, we assign $T(S) = 1$, so that these holes are always retained in the filtering process. Many other types of increasing criteria are possible—see Reference [75] for some of these examples.

After defining holes, filling of holes, and increasing criteria, we define an *inclusion sequence* to complete the requirement for formally defining an inclusion filter. Let L be a binary image and $\mathbf{x}$ be a pixel in it. Two cases are possible. Case I: $L(\mathbf{x}) = 1$ (i.e., $\mathbf{x}$ is a foreground pixel). In this case, in the adjacency forest for L, we will have a sequence of connected components $C_1, H_2, C_3, ...,$ so that $\mathbf{x} \in C_1$ and C_1 is a child of H_2, H_2 is a child of C_3, and so on. We denote foreground-connected components by C's and background-connected components by H's. For case II, $L(\mathbf{x}) = 0$, i.e., $\mathbf{x}$ is a background-connected component. In this case, a similar sequence: H_1, C_2, H_3, ... will exist. For a pixel $\mathbf{x}$, we call this sequence an *inclusion sequence*.

Let H_1, C_2, ..., (or C_1, H_2, ...) be the inclusion sequence for a pixel $\mathbf{x}$ in the binary image L. Let T be an increasing criterion. We can now define the result of the inclusion filter on L as:

$$L_f(\mathbf{x}) = \begin{cases} 1, & \text{if } T(F(H_n)) = 0 \text{ and } T(F(C_{n+1})) = 1, \\ 0, & \text{if } T(F(C_n)) = 0 \text{ and } T(F(H_{n+1})) = 1, \end{cases} \qquad (6.22)$$

for some $n \geq 0$. In this context, $F(H_0)$ or $F(C_0)$ is an empty set, for which we assign T as 0. We also denote: $L_f(\mathbf{x}) = \Psi_T(L(\mathbf{x}))$, where Ψ_T is the underlying inclusion filter with an increasing criterion T.

So far, we have defined inclusion filter for a binary image. Since binary inclusion filters are increasing [75], extension of the binary inclusion filter to a grayscale image is performed via a *stacking algorithm* as follows:

1. Decompose grayscale image $I(\mathbf{x})$ into binary images:

    $$L^\lambda(\mathbf{x}) = \begin{cases} 1, & \text{if } I(\mathbf{x}) \geq \lambda, \\ 0, & \text{otherwise.} \end{cases}$$

2. Perform inclusion filter on each binary image:

    $$L_f^\lambda(\mathbf{x}) = \Psi_T(L^\lambda(\mathbf{x})).$$

3. Stack-filtered binary images $L_f^\lambda(\mathbf{x})$:

    $$L_f^\lambda(\mathbf{x}) : I_f(\mathbf{x}) = \max\{\lambda : L_f^\lambda(\mathbf{x}) = 1\}.$$

When we apply this marker filter to the MRI of a lung shown in Figure 6.6a, we obtain the result shown in Figure 6.6b. The marker points are shown as bright dots in Figure 6.6a. Canny edge detection results of these images are shown, respectively, in Figure 6.6c and 6.6d. Notice that inclusion filter has helped homogenize the lung so that contrasting lung boundaries are salient in Figure 6.6d. For this example, one can argue in favor of the marker inclusion filter in comparison to, say, area-open-/area-close-connected operator in the following manner. Because the lung cavity size varies widely in size among different MRI slices, it may be difficult to fix the area threshold even for a single patient. However, finding two sets of marker points that stays within the lung cavity in all MRI slices may actually be possible. Thus, the motivation for inclusion filter, especially marker

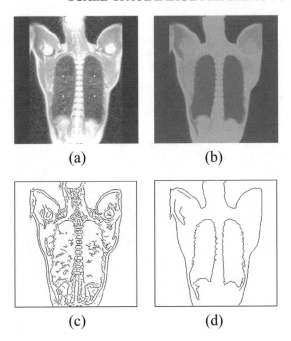

FIGURE 6.6: (a) An MR image slice with two marker points inside the lung cavity. (b) Marker inclusion filter results on (a). (c) Canny edge detection on (a). (d) Canny edge detection on (b) [75].

inclusion filter, is to homogenize (smooth) the area within the image that contains these markers. Such filters can easily deal with salt and pepper noise, as well as unwanted clutter not containing the marker points.

In summary, this chapter has exhibited scale-space filters that improve segmentation of biomedical images. Although these filters have value in enhancement, their ability to improve the partitioning of an image into meaningful regions has been emphasized.

•   •   •   •

Acknowledgments

There is a new breed of engineer that attempts to know both worlds—that of engineering and that of biology and medicine. The engineers that wrote this book are old-fashioned in the sense that they believe that excellence with any degree of depth cannot be achieved in both domains by the same individual. We believe that our biomedical image analysis work falls flat without strong collaborators. Through the years, we have been blessed with several. The laboratory has enjoyed a long and productive relationship with Dr. Klaus Ley of the La Jolla Institute. Thanks to Klaus, you will see several images containing leukocytes in this book. In this area, we have also worked with Dr. Ed Damiano of BU and his novel particle image velocimetry data. Cardiac (and prostate) collaborators that contributed to the data shown in this book include Dr. John Hossack (ultrasound), Dr. Brent French (magnetic resonance imaging [MRI]/ultrasound/microscopy), and Dr. Fred Epstein (MRI), all of the University of Virginia. Dr. Steve Millington of Vienna and Dr. Shep Hurwitz worked with us on orthopedic images. More recently, we have joined with Dr. Marios Pattichis of the University of New Mexico and Dr. Peter Soliz of VisionQuest-Biomedical on the subject of retinal images.

In terms of contributors to our segmentation work, Dr. Dipti Prasad Mukherjee of the Indian Statistical Institute has been a long-time collaborator with the Virginia Image and Video Analysis (VIVA) laboratory. Many of VIVA's students contributed to work highlighted in this book including Dr. Joseph Bosworth, Dr. Yongjian Yu, Dr. Bing Li, Dr. Gang Dong, and Dr. Andrew Gilliam. Dr. Yu and Dr. Mukherjee, as well as current VIVA students, Saurav Basu and Josh Poulin, served as reviewers for this book.

We also thank the National Institutes of Health and recognize our funding and that of our collaborators under the following grants: HL68510, HL082870, HL082870, EB001826, EB001763. We thank the National Science Foundation, the Army Research Office for support under 46850-CI, and NSERC, Canada.

Lastly, we are grateful for the involvement of editor Al Bovik and publisher Joel Claypool.

References

[1]   M. Kass, A. Witkin, and D. Terzopoulos, "Snakes: Active contour models," *International Journal of Computer Vision*, pp. 321–331, 1988. doi:10.1007/BF00133570

[2]   M. A. Gennert and A. Y. Yuille, "Determining the optimal weights in multiple objective function optimization," in *Proceedings, 2nd International Conference on Computer Vision*, pp. 87–89, 1988. doi:10.1109/CCV.1988.589974

[3]   J. L. Troutman, *Variational calculus with elementary convexity*, New York: Springer-Verlag, 1983.

[4]   C. A. Hall and T. A. Porsching, *Numerical analysis of partial differential equations*, Prentice Hall: Englewood Cliffs, NJ, 1990.

[5]   L. D. Cohen, "On active contour models and balloons." *CVGIP: Image Understanding*, vol. 53, pp. 211–218, 1991. doi:10.1016/1049-9660(91)90028-N

[6]   C. Xu and J. L. Prince, "Snakes, shapes, and gradient vector flow," *IEEE Transactions on Image Processing*, vol. 7, pp. 359, 1998.

[7]   N. Ray, S. T. Acton, T. Altes, E. E. de Lange, and J. R. Brookeman, "Merging parametric active contours within homogeneous image regions for MRI-based lung segmentation," *IEEE Transactions on Medical Imaging*, vol. 22, no. 1, pp. 189–199, 2003. doi:10.1109/TMI.2002.808354

[8]   B. Li and S. T. Acton, "Active contour external force using vector field convolution for image segmentation," *IEEE Transactions on Image Processing*, vol. 16, pp. 2096–2097, 2007.

[9]   X. Ge, X. Ge, and J. Tian, "An automatic active contour model for multiple objects," in *Proceedings of the International Conference on Pattern Recognition*, 2002. doi:10.1109/ICPR.2002.1048444

[10] C. Li, J. Liu, and M. D. Fox, "Segmentation of external force field for automatic initialization and splitting of snakes," *Pattern Recognition*, vol. 38, pp. 1947, 2005.

[11] B. Li and S. T. Acton, "Automatic active model initialization via Poisson inverse gradient," *IEEE Transactions on Image Processing*, vol. 17, pp. 1406–1420, 2008.

[12] Y. Saad, *Iterative methods for sparse linear systems*, 2nd ed: Society for Industrial and Applied Mathematics, 2003.

[13]   K. A. Atkinson, *An introduction to numerical analysis*, 2nd ed: John Wiley and Sons, 1988.

[14]   L. Giraud, R. Guivarch, and J. Stein, "Parallel distributed FFT-based solvers for 3-D Poisson problems in meso-scale atmospheric simulations," *International Journal of High Performance Computing Applications*, vol. 15, pp. 36–46, 2001. doi:10.1177/109434200101500104

[15]   T. Matsumoto and T. Hanawa, "A fast algorithm for solving the Poisson equation on a nested grid," *The Astrophysical Journal*, vol. 583, pp. 296–307, 2003. doi:10.1086/345338

[16]   B. Li and S. T. Acton, "Feature weighted active contours," *Proceedings, IEEE Southwest Symposium on Image Analysis and Interpretation*, Denver, Colorado, March 26–28, 2006.

[17]   J. H. Ahlberg, E. N. Nilson, and J. L. Wash, *The theory of splines and their applications*, New York: Academic, 1967.

[18]   J. Gielis, "A generic geometric transformation that unifies a wide range of natural and abstract shapes," *American Journal of Botany*, vol. 90, pp. 333–338, 2003.

[19]   G. Dong, N. Ray, and S. T. Acton, "Intravital leukocyte detection using the gradient inverse coefficient of variation," *IEEE Transactions on Medical Imaging*, vol. 24, no. 7, pp. 910–924, July 2005.

[20]   S. Sahoo, N. Ray, and S. T. Acton, "Rolling leukocyte detection based on teardrop shape and the gradient inverse coefficient of variation," *International Conference on Medical Information Visualisation—BioMedical Visualisation 2006*, pp. 29–33, July 5–7, 2006. doi:10.1109/MEDIVIS.2006.22

[21]   L. D. Cohen and I. Cohen, "Finite-element methods for active contour models and balloons for 2-D and 3-D images," *IEEE Transactions on Pattern Analysis and Machine Intelligence*, vol. 15, pp. 1131–1147, 1993.

[22]   W. Press, S. Teukolsky, W. Vetterling, and B. Flannery, *Numerical Recipes in C*, 2nd ed. Cambridge, MA: Cambridge University Press, 1992.

[23]   K. S. Arun, T. S. Huang, and S. D. Blostein, "Least-squares fitting of two 3-D point sets," *IEEE Transactions on Pattern Analysis and Machine Intelligence*, vol. 9, pp. 698–700, 1987.

[24]   S. Umeyama, "Least-squares estimation of transformation parameters between two point patterns," *IEEE Transactions on Pattern Analysis and Machine Intelligence*, vol. 13, pp. 376–380, 1991. doi:10.1109/34.88573

[25]   B. Li, S. A. Millington, D. D. Anderson, and S. T. Acton, "Registration of surfaces to 3D images using rigid body surfaces," *Asilomar Conference on Signals, Systems and Computers*, Pacific Grove, California, October 29–November 1, 2006.

[26]   R. Janiczek, N. Ray, F. Epstein, and S. T. Acton, "A Markov chain Monte Carlo method for tracking myocardial borders," invited paper at *IS&T/SPIE's 17th Annual Symposium on Electronic Imaging Science and Technology*, January 16–20, 2005. doi:10.1117/12.598864

[27] S. T. Acton and N. Ray, *Biomedical Image Analysis: Tracking*, A. C. Bovik, editor, Morgan and Claypool Publishers, 2006.

[28] O. Häggström, Finite Markov chains and algorithmic applications, Cambridge University Press, Cambridge, 2002.

[29] T. F. Cootes, C. J. Taylor, D. H. Cooper, and J. Graham, "Active shape models—their training and applications," *Computer Vision and Image Understanding*, vol. 61, no. 1, pp. 38–59, 1995. doi:10.1006/cviu.1995.1004

[30] Z. Xue, S. Z. Li, and E. K. Teoh, "Ai-eigensnake: An affine-invariant deformable contour model for object matching," *Image and Vision Computing*, vol. 20, no. 2, pp. 77–84, 2002. doi:10.1016/S0262-8856(01)00078-6

[31] Z. Zhang, "Iterative point matching for registration of free-form curves and surfaces," *International Journal of Computer Vision*, vol. 13, no. 2, pp. 119–152, 1994. doi:10.1007/BF01427149

[32] J. E. Pickard, S. T. Acton, and J. A. Hossack, "The effect of initialization and registration on the active shape segmentation of the myocardium in contrast enhanced ultrasound," *Proceedings, 2005 IEEE International Ultrasonics Symposium*, Rotterdam, The Netherlands, September 18–21, 2005. doi:10.1109/ULTSYM.2005.1603286

[33] K. F. Lai and R. T. Chin, "Deformable contours: modeling and extraction," *IEEE Transactions on Pattern Analysis and Machine Intelligence*, vol. 17, no. 11, pp. 1084–1090, 1995. doi:10.1109/34.473235

[34] K. Lai and R. Chin, "On regularization, formulation and initialization of the active contour models (snakes)," *Asian Conference on Computer Vision*, pp. 542–545, 1993.

[35] S. Osher and L. Rudin, "Feature-oriented image enhancement using shock filters," *SIAM Journal on Numerical Analysis*, vol. 27, pp. 919–940, 1990. doi:10.1137/0727053

[36] J. A. Sethian, Level set methods and fast matching methods, Cambridge University Press, 1999.

[37] T. F. Chan and L. Vese, "Active contours without edges," *IEEE Transactions on Image Processing*, vol. 10, pp. 266–277, 2001. doi:10.1109/83.902291

[38] A. Yezzi, J. A. Tsai, and A. Willsky, "A statistical approach to snakes for bimodal and trimodal imagery," *Proceedings of ICCV*, pp. 898–903, September, 1999.

[39] S. Osher, "Riemann solvers, the entropy condition, and difference approximations," *SIAM Journal on Numerical Analysis*, vol. 21, pp. 217–235, 1984. doi:10.1137/0721016

[40] A. Yezzi, S. Kichenassamy, A. Kumar, P. Olver, and A. Tannenbaum, "A geometric snake model for segmentation of medical imagery," *IEEE Transactions on Medical Imaging*, vol. 16, pp. 199–209, 1997. doi:10.1109/42.563665

[41]   Y. Yu and S. T. Acton, "Active contours with area-weighted binary flows for segmenting low SNR imagery," *Proceedings, IEEE International Conference on Image Processing*, Barcelona, September 14–17, 2003.

[42]   T. F. Chan and L. A. Vese, "Active Contours Without Edges," *IEEE Transactions on Image Processing*, vol. 10, pp. 266–277, 2001.

[43]   F. H. Y. Chan, F. K. Lam, and H. Zhu, "Adaptive thresholding by variational method." *IEEE Transactions on Image Processing*, vol. 7, no. 3, pp. 468–473, 1998. doi:10.1109/83.661196

[44]   S. D. Yanowitz and A. M. Bruckstein, "A new method for image segmentation," *Computer Vision, Graphics, and Image Processing*, vol. 46, no. 1, pp. 82–95, 1989. doi:10.1016/S0734-189X(89)80017-9

[45]   G. Dong and S. T. Acton, "A variational method for leukocyte detection," *Proceedings, IEEE International Conference on Image Processing*, Barcelona, September 14–17, 2003.

[46]   D. Mumford and J. Shah, "Boundary detection by minimizing functionals," in *Proceedings, IEEE Conference on Computer Vision and Pattern Recognition*, San Francisco, CA, 1985.

[47]   G. A. Hewer, C. Kenney, and B. S. Manjunath, "Variational image segmentation using boundary functions," *IEEE Transactions on Image Processing*, vol. 7, pp. 1269–1282, 1998. doi:10.1109/83.709660

[48]   A.V. Goldberg and C. Harrelson, "Computing the shortest path: A* search meets Graph Theory," in *Proceedings, 16th ACM-SIAM Symposium on Discrete Algorithms*, pp. 156–165, 2005.

[49]   T. H. Cormen, C. E. Leiserson, R. L. Rivest, and C. Stein, *Introduction to Algorithms*, MIT Press, 2001.

[50]   V. Kolmogorov and R. Zabih, "What energy functions can be minimized via graph cuts?" *IEEE Transactions on Pattern Analysis and Machine Intelligence*, vol. 26, no. 2, pp. 147–159, 2004. doi:10.1109/TPAMI.2004.1262177

[51]   J. Shi and J. Malik, "Normalized cuts and image segmentation," *IEEE Transactions on Pattern Analysis and Machine Intelligence*, vol. 22, no. 8, pp. 888–905, 2000.

[52]   D. M. Greig, B. T. Porteous, and A. H. Seheult, "Exact maximum a posteriori estimation for binary images," *Journal of Royal Statistical Society B*, vol. 51, no. 2, pp. 271–279, 1989.

[53]   M. R. Garey and D. S. Johnson, *Computers and Intractability: A guide to the theory of NP-completeness*, W.H. Freeman and Company, New York, 1979.

[54]   N. Otsu, "A threshold selection method from gray-level histogram," *IEEE Transactions on System Man Cybernatics*, vol. SMC-9, no. 1, pp. 62–66, 1979.

[55]   T. Hastie, R. Tibshirani, and J. Friedman, *The Elements of Statistical Learning: Data Mining, Inference, and Prediction*, Springer, 2001.

[56] J. H. Bosworth and S. T. Acton, "Morphological scale-space in image processing," *Digital Signal Processing*, vol. 13, pp. 338–367, 2003. doi:10.1016/S1051-2004(02)00033-7

[57] A. C. Bovik and S. T. Acton, "Basic linear filtering with application to image enhancement," in the *Handbook of Image and Video Processing*, Second Edition, Academic Press, 2004.

[58] J. Morel and S. Solimini, *Variational Methods in Image Segmentation*, Birkhauser, Boston, 1995.

[59] P. Perona and J. Malik, "Scale-space and edge detection using anisotropic diffusion," *IEEE Transactions on Pattern Analysis and Machine Intelligence*, vol. 12, 629–639, 1990. doi:10.1109/34.56205

[60] A. P. Witkin, Scale-space filtering. In *Proceedings, International Joint Conference on Artificial Intelligence*, Palto Alto, CA, 1983, pp. 1019–1022.

[61] S. T. Acton, "Multigrid anisotropic diffusion," *IEEE Transactions on Image Processing*, vol. 7, pp. 280–291, 1998. doi:10.1109/83.661178

[62] F. Catte, P.-L. Lions, J.-M. Morel, and T. Coll, "Image selective smoothing and edge detection by nonlinear diffusion," *SIAM Journal on Numerical Analysis*, vol. 29, pp. 182–193, 1992. doi:10.1137/0729012

[63] L. Alvarez, P.-L. Lions, and J.-M. Morel, "Image selective smoothing and edge detection by nonlinear diffusion II," *SIAM Journal on Numerical Analysis*, vol. 29, pp. 845–866, 1992. doi:10.1137/0729052

[64] A. El-Fallah and G. Ford, "The evolution of mean curvature in image filtering," *Proceedings, IEEE International Conference on Image Processing*, Austin, Texas, Nov. 1994.

[65] J. S. Lee, "Digital image enhancement and noise filtering by using local statistics," *IEEE Transactions on Pattern Analysis and Machine Intelligence*, vol. PAM1-2, pp. 165–168, 1980.

[66] Y. Yu and S. T. Acton, "Speckle reducing anisotropic diffusion," *IEEE Transactions on Image Processing*, vol. 11, pp. 1260–1270, 2002. doi:10.1109/TIP.2004.836166

[67] Y. Yu and S. T. Acton, "Edge detection in ultrasound imagery using the instantaneous coefficient of variation," *IEEE Transactions on Image Processing*, vol. 13, pp. 1640–1655, 2004.

[68] Q. Sun, J. Hossack, J. Tang, and S. T. Acton, "Speckle reducing anisotropic diffusion for 3-D ultrasound images," *Computerized Medical Imaging and Graphics*, vol. 8, pp. 461–470, 2004.

[69] A. Restrepo (Palacios) and A. C. Bovik, "Locally monotonic regression," *IEEE Transactions on Signal Processing*, vol. 41, pp. 2796–2810, 1993.

[70] A. Restrepo (Palacios) and S. T. Acton, "2-D binary locally monotonic regression," *Proceedings, IEEE International Conference on Acoustics, Speech and Signal Processing (ICASSP-99)*, Phoenix, March 14–19, 1999.

[71] S. T. Acton, "Locally monotonic diffusion," *IEEE Transactions on Signal Processing*, vol. 48, pp. 1379–1389, 2000.

[72] S. T. Acton and A. C. Bovik, "Segmentation by locally monotonic reduction," *Journal of Applied Signal Processing*, vol. 6, pp. 42–54, 1999. doi:10.1007/s005290050033

[73] S. T. Acton, "Fast algorithms for area morphology," *Digital Signal Processing*, vol. 11, pp. 187–203, 2001. doi:10.1006/dspr.2001.0386

[74] P. Salembier and J. Serra, "Flat zones filtering, connected operators and filters by reconstruction," *IEEE Transactions on Image Processing*, vol. 4, pp. 1153–1160, August 1995. doi:10.1109/83.403422

[75] N. Ray and S. T. Acton, "Inclusion filters: A class of self-dual connected operators," *IEEE Transactions on Image Processing*, vol. 14, no. 11, pp. 1736–1746, Nov. 2005.

[76] S. Millington, B. Li, J. Tang, S. Trattnig, J. R. Crandall, S. R. Hurwitz, and S. T. Acton, "Quantitative and topographical evaluation of ankle articular cartilage using high resolution MRI," *Journal of Orthopaedic Research*, vol. 25, pp. 143–151, 2006 (referenced in figures).

[77] S. T. Acton and K. Ley, "Tracking leukocytes from *in vivo* video microscopy using morphological anisotropic diffusion," *Proceedings, IEEE International Conference on Image Processing*, Thessaloniki, Greece, October 7–10, 2001.

Author Biographies

Scott T. Acton received his Ph.D. degree in electrical and computer engineering from the University of Texas at Austin, where he was a student of Al Bovik. He was the class of 1984 valedictorian at Oakton High School in Vienna, Virginia. Prof. Acton has worked in industry for AT&T, Oakton, VA, the MITRE Corporation, McLean, VA, and Motorola, Inc., Phoenix, AZ, and in Academia for Oklahoma State University, Stillwater. Currently, he is a professor at the University of Virginia (UVa), where he is a member of the Charles L. Brown Department of Electrical and Computer Engineering and the Department of Biomedical Engineering. Prof. Acton was selected as the Eta Kappa Nu Outstanding Young Electrical Engineer—a national award that has been given annually since 1936. At UVa, he was named the Walter N. Munster Chair for Intelligence Enhancement. Prof. Acton is an active participant in the IEEE, serving as associate editor for the *IEEE Transactions on Image Processing*. He was the 2004 Technical Program Chair and the 2006 General Chair for the *Asilomar Conference on Signals, Systems, and Computers*. His research interests include active models, biomedical segmentation problems, biomedical tracking problems, content-based image retrieval, and partial differential equations. Prof. Acton lives in beautiful Charlottesville, VA, and enjoys hiking, basketball, running, golf, and chasing his two boys. He spent the 2007–2008 academic year in Santa Fe, NM.

Nilanjan Ray received his bachelor's degree in mechanical engineering from Jadavpur University, Calcutta, India, in 1995; his master's degree in computer science from Indian Statistical Institute, Calcutta, in 1997; and his Ph.D. in electrical engineering from the University of Virginia, Charlottesville, in 2003. After postdoctoral and industrial work experience, he is now an assistant professor at the Department of Computing Science, University of Alberta, Canada. His research area is image analysis.

Printed in the United States
by Baker & Taylor Publisher Services